S0-BOE-523

ATLAS
OF THE
LIVING WORLD

Contents

CIP data
Colombo, Federica
Atlas of the living world.
1. Animal ecology
I. Title   II. Atlante del mondo vivente.
English
591. 5   QH541

ISBN 0 222 00870 9

First published in the English language 1983
Reprinted 1984
© Burke Publishing Company Limited 1983
Translated and adapted from "Atlante del Mondo Vivente"
© Vallardi Industrie Grafiche S.p.A. 1980

ISBN 0 222 008870 9

Burke Publishing Company Limited
Pegasus House, 116-120 Golden Lane, London EC1Y OTL, England.
Burke Publishing (Canada) Limited
Registered Office: 20 Queen Street West, Suite 3000, Box 30,
Toronto, Canada M5H 1V5
Burke Publishing Company Inc.
Registered Office: 333 State Street, Po Box 1740 Bridgeport,
Connecticut 06601, U.S.A.
Printed by Vallardi Industrie Grafiche S.p.A. Milan.

ACKNOWLEDGEMENTS

The publishers are grateful to Professor Robert D. Martin F.I. Biol. and Anne-Elise Martin for assistance in the preparation of the text for this English-language edition.

The drawings reproduced in this book are by Gabriele Pozzi and Fernando Russo.

The maps are reproduced by kind permission of Vallardi Industrie Grafiche.

The photographs are reproduced by kind permission of: Coleman; Dimt; Fiore; Fratas; Gaggero; Jacana; Mazza; Mairani; Marka; NASA; Pagani; Pedone; Prato-Previde; Publiaerfoto; Regaldi; Ricatto; Sterna; Stutte; Tiofoto; Vallardi and Zappelli.

ATLAS
OF THE
LIVING WORLD

Burke Books ▶️B   LONDON * TORONTO * NEW YORK

28486
28483

North Pole

ARCTIC OCEAN

Ellesmere I.
Queen Elizabeth Is.
Magnetic Pole
Greenland

Banks I.
Beaufort Sea
Baffin Bay
Baffin I.
Davis Strait

C. Barrow
Victoria I.
Boothia Pen.
Melville Pen.

ASIA
Brooks Range
Great Bear L.
Mackenzie
Great Slave L.
Hudson Strait
C. Farewell

Bering Strait
Yukon
Mt. McKinley
Mt. Logan
Hudson Bay
Ungava Pen.
Labrador

Nunivak I.
Bering Sea
Gulf of Alaska
Alexander Arch.
Coast Range
L. Winnipeg
Belcher Is.

Aleutian Is.
7860
Aleutian Trench
Vancouver I.

ROCKY MOUNTAINS
Belle Isle Str.
Newfoundland
St. Lawrence
L. Superior

NORTH AMERICA

L. Huron
L. Michigan
Mississippi
L. Ontario
St. Lawrence
L. Erie
C. Sable

6741
Great Plains
Missouri
Mitchell
APPALACHIAN MTS.
C. Cod

C. Mendocino
Pikes Peak
Colorado
Arkansas
Red R.
C. Hatteras

Whitney
Sierra Nevada
Rio Grande
Bermuda

Guadalupe I.
G. of California
Western Sierra Madre
Mexican Plateau
Eastern Sierra Madre
Florida
Sargasso Sea
6996

Hawaiian Islands
TROPIC OF CANCER
Lower California
Bahamas
Greater Antilles

Midway I.
C. San Lucas
Gulf of Mexico
Cuba
Puerto Rico
6225

Johnston I.
Hawaii
Revilla Gigedo Is.
G. of Campeche
Yucatan
Jamaica
Hispaniola
Puerto Rico
Guadeloupe

Citlaltepetl
G. of Honduras
Caribbean Sea
Martinique
Lesser Antilles
Barbados

PACIFIC
Isthmus of Tehuantepec
Tajumulco
Trinidad

Nicaragua
Pta. Gallinas

Clipperton I.
Isthmus of Panama
Lake Maracaibo
Orinoco

Palmyra I.
G. of Panama
Chirripó
Guiana Highlands

Baker Is.
7251
Fanning I.
Christmas I.
Galápagos Is.
Gulf of Guayaquil
Huila
Llanos
Japurá
Amazon
Marajó I.

EQUATOR
LONGITUDE WEST FROM GREENWICH
Chimborazo
Amazon
Fernando de Noronha I.

Phoenix Is.
Malden I.
Starbuck I.
Caroline I.
Huascarán
6768
Selvas
Madeira
Purus
C. S. Roque

Tokelau Is.
Marquesas Is.
SOUTH
Sertões
Ascension

W. Samoa
OCEAN
AMERICA
Caatingas
São Francisco

Tonga Is.
(Friendly Is.)
Society Is.
Tahiti
Tuamotu Arch.
Illampu
Plateau of Mato Grosso
Brazilian Highlands

Tonga Trench
Cook Is.
Tubuai Is.
L. Titicaca
Bolivian Plat.
Pico da Bandeira

10882
TROPIC OF CAPRICORN
ANDES
Gran Chaco
Paraguay
Paraná
C. Frio

Kermadec Trench
Rapa
Pitcairn I.
Ducie I.
Sala-y-Gómez
Easter I.
Aconcagua
Pampas
Paraná
Lagoa dos Patos

10047
Kermadec Is.
S. Félix
S. Ambrosio
MOUNTAINS
Rio de la Plata

Juan Fernández Is.
Pampas

Chatham Is.
Chiloé
Patagonia
G. of San Matías

Chonos Arch.
G. of San Jorge

Str. of Magellan
Falkland Is.
South Georgia

Tierra del Fuego
8264

C. Horn
Drake Passage
South Sandwich Is.

ANTARCTIC CIRCLE
South Shetland Is.
South Orkney Is.

Ross Sea
Antarctic Peninsula

Byrd Land
Charcot I.
Alexander I.
Weddell Sea

Ellsworth L.
Coats Land

South Pole

ATLANTIC
Azores
C. St. Vincent
Madeira
Canary Is.
C. Blanco
Mauritania
C. Verde Is.
C. Verde
C. Palmas
Ascension
Tristan da Cunha
Gough

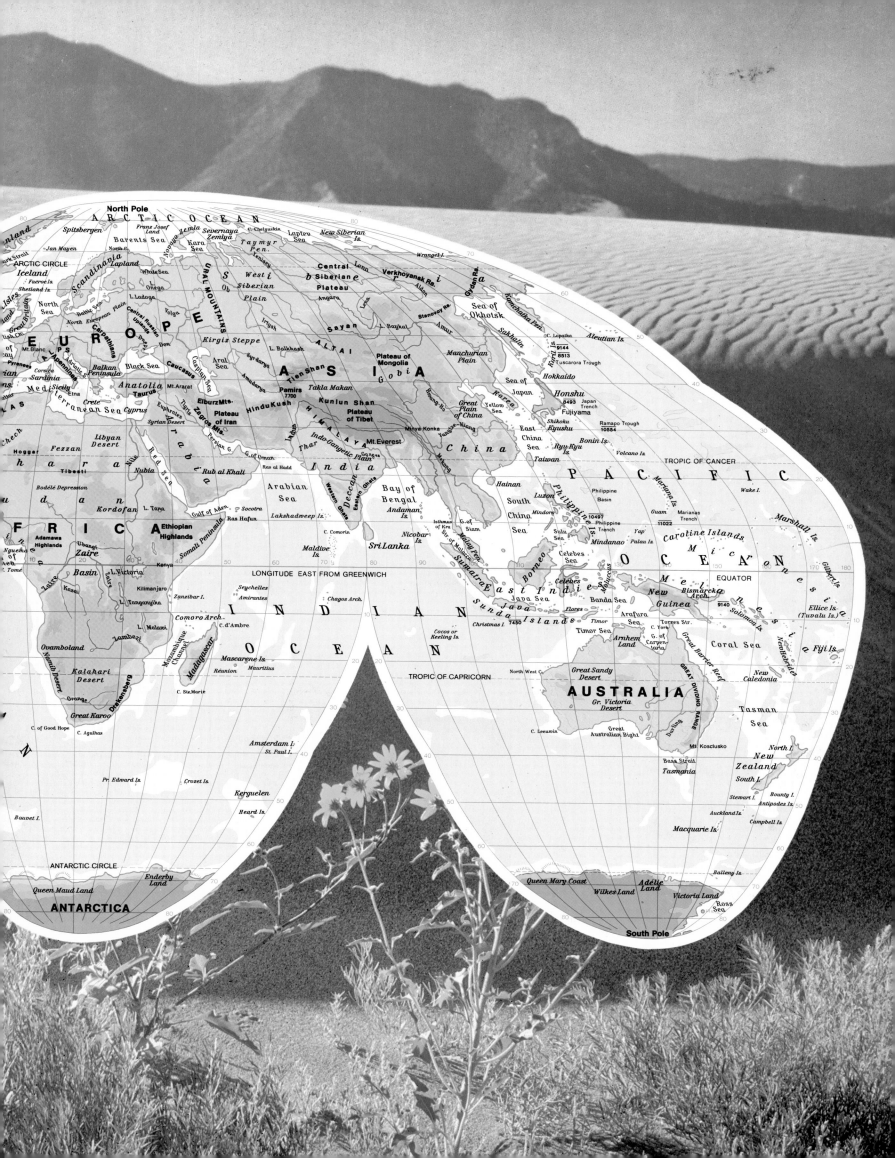

Planet Earth

The more we discover about the Earth through scientific study, the more it becomes obvious that our planet is quite exceptional in comparison with its neighbours in the solar system. Various space probes which have been guided onto the nearest planets have transmitted back pictures showing just how monotonous their landscapes are; and it is clear from the physical and chemical characteristics of these planets that they cannot support any form of life as we understand it. The distance of the Earth from the Sun, combined with its complex pattern of rotating on its axis while orbiting the Sun, ensures the presence of equable temperatures. These barely permit the freezing of water under certain conditions yet remain well below its boiling-point. The high spe-cific gravity of the Earth (5.5) is combined with only moderate dimensions (equatorial diameter: less than 13,000 kilometres or 8,000 miles), which accounts for the presence of a relatively thick atmospheric layer composed primarily of nitrogen (78%) and oxygen (20%), unique within our solar system. In addition, water is present in great abundance and not only plays an essential role in the dynamic change of the Earth's surface over a period of time, but also provides one of the indispensable conditions for life. Downhill torrents wear away softer material to expose isolated rocky outcrops, while glaciers have a pronounced influence on both erosion and the transport of material in mountain-chains.

Structural Characteristics of the Earth

The internal structure of the Earth remains a fascinating subject for research, with many mysteries still to be resolved. Despite the prodigious progress which has been made in various branches of science, man is still unable to penetrate directly into the Earth's core. The deepest drillings so far made have extended no more than ten kilometres (six miles), which is very little in comparison with the terrestrial radius of 6,000 kilometres (4,000 miles). Nevertheless, there is information available from a variety of research areas—including geophysics, geochemistry, astronomy and geology—which strongly suggests that the Earth consists of a series of concentric layers. For example, there is a marked discrepancy between the average specific gravity of the Earth (5.5) and that of the solid crust (3.3). This suggests that the crust is much less dense than the underlying matter. The most interesting information about the Earth's interior has been obtained from examination of sound (seismic) wave evidence. Analysis of seismographic recordings has shown that the speed of propagation of seismic waves generated by earthquakes depends upon the density of the rocks through which they pass. The numerous layers which make up the Earth can hence be grouped into three major zones: the outermost crust, the intervening mantle and the central core. The boundaries between these three zones (major discontinuities) are clearly marked by changes in density and other physical properties of the masses through which seismic waves pass.

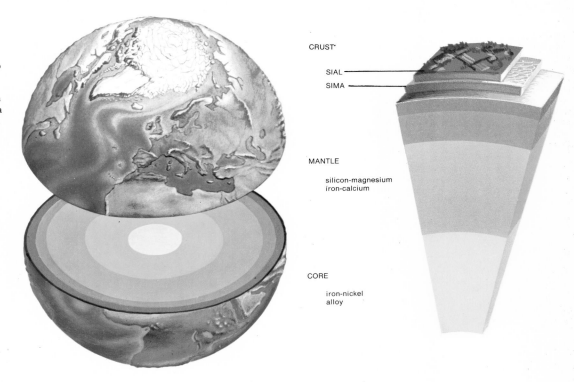

CRUST

SIAL

SIMA

MANTLE

silicon-magnesium
iron-calcium

CORE

iron-nickel
alloy

Terrestrial Magnetism

The Earth behaves like a gigantic magnet whose poles are oriented more or less in line with the geographical poles. The origin of terrestrial magnetism has yet to be fully explained, though it is generally believed that it is related in some way to the Earth's rotation. In fact, the semi-liquid mantle, along with the crust at its surface, probably rotates more quickly than the supposedly solid core. This results in the formation of a kind of "natural dynamo" generating a magnetic field. However, there is now reason to believe that some influence may also be exerted by external magnetic forces of stellar and solar origin.

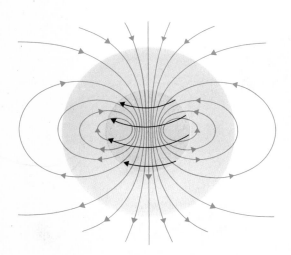

The Crust and the Mantle

The upper part of the crust—the part found in continental regions—is constituted by so-called "sial" (lighter rocks formed mainly by silicon and aluminium), whereas the ocean floors consist of "sima" (heavier rocks composed of silicon and magnesium). The thickness of the Earth's crust varies from thirty kilometres (seventeen miles) in continental regions to five kilometres (three miles) on the ocean floor. Beneath the crust lies the mantle, and the two layers are clearly separated by the Mohorovicic discontinuity (named after the geophysicist who discovered it). The upper layer of the mantle and the crust together form the lithosphere, while the lower zone of the mantle (referred to as the asthenosphere) is almost fluid because of the special temperature and pressure conditions encountered at that level.

Regions of Seismic and Volcanic Activity

Earthquakes and volcanoes together provide the most obvious evidence of the instability of the Earth's crust. For some time, volcanic and seismic activity were seen as marginal products of dynamic change in the Earth's crust, but they are now seen as an integral part of the modern theory of plate tectonics. The map below shows the close agreement in geographical location between volcanoes and zones of seismic activity. Volcanoes are primarily found along the edges of continental plates, where there is either creation of new crust (oceanic ridges) or absorption and destruction of existing crust material (subduction trenches). Seismic activity reflects both dislocation between masses of rock and friction in zones of contact between plates. Seismic waves generated along oceanic trenches, indicating that one plate is being forced down beneath another, are particularly intense.

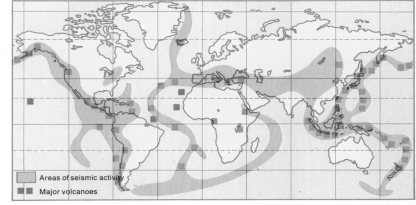

DISTRIBUTION OF AREAS OF VOLCANIC AND SEISMIC ACTIVITY

Areas of seismic activity

Major volcanoes

Erosion

The modern contours of the Earth's landscape are not the direct product of internal processes (mountain-building, volcanic activity, etc.). The Earth's crust has, instead, been modelled by external influences which have combined to sculpture the varied landscapes seen at different latitudes. Erosion, which involves a complex series of physical and chemical processes, tends to wear down anything that has been raised up by endogenous activity (originating internally). Water, in its various physical states, is very active in this respect. (*Above:* Erosion pyramids in Cappadocia. *Below:* The Rhône glacier.)

Continental Drift and Plate Tectonics

It is by no means a new idea that the continents can move about over the surface of the Earth and that they once formed a single super-continent. A simple glance at a geographical map is sufficient to see that the present-day continents could easily be fitted together, particularly in the case of Africa and South America. However, the first coherent theory of continental drift was put forward only at the beginning of the twentieth century. Alfred Wegener's *The Origin of Continents and Oceans* was published in 1915. In this book, the German scientist brought together a diverse array of relevant facts from geology, geophysics and palaeontology. He postulated the original existence of a single gigantic continent, Pangea, consisting of solid but relatively light-weight sialic rock and superimposed upon a denser, but more easily deformed, layer of sima. Pangea became fragmented to form the modern continental land-masses less than a hundred million years ago. These land-masses drifted over the surface of the Earth and occasionally collided with one another to give rise to orogenesis (formation of mountain-chains). Proceeding from the matching of the contours of our present-day continents, Wegener then assembled evidence of complementary geological formations and fossil deposits on continental margins now separated by oceans. Wegener's ideas provoked a long controversy which focused on the weakest point in his theory—the necessity to demonstrate the existence of the immense forces required to shift the continental land-masses. It was not until the 1960s that irrefutable evidence was obtained to justify acceptance of Wegener's theory of continental drift in a somewhat modified form. Collaborative work between geologists and geophysicists has now yielded an acceptable theory of plate tectonics which combines in one all-embracing framework all the major phenomena of the Earth's surface: orogenesis, volcanic activity, seismic waves, deep-sea trenches and oceanic ridges. According to this unified theory, the Earth's crust and the upper layer of the mantle (lithosphere) are fragmented into a number of plates which are continually moving over the asthenosphere in an ever-changing pattern. The motor force behind these movements is provided by convection currents in the asthenosphere, arising because of temperature differences. Seven principal continental plates are now recognized: European, North American, South American, Antarctic, Indo-Australian, North Pacific and South Pacific.

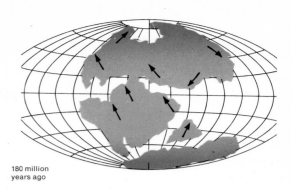

180 million
years ago

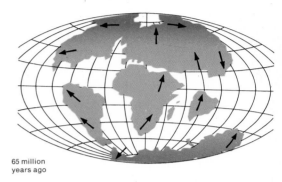

65 million
years ago

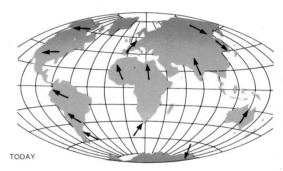

TODAY

Movement of the continental plates

Formation and Destruction of the Crust

All the major events involving the Earth's crust occur along the boundaries of the plates. It is at these boundaries that new crust is generated and ancient crust is destroyed. The former takes place in zones of sea-floor spreading which are characterized by fracture zones (oceanic ridges; rift valleys) along which magma escapes to the surface and solidifies to form new crust material. The new material exerts pressure along the boundaries of the fracture zones, thus giving rise to spreading of the ocean floor. In subduction zones, on the other hand, destruction of the crust takes place. Such zones consist of deep-sea trenches which may be as much as 10,000 metres (33,000 feet) in depth. At each subduction trench, one tectonic plate is forced down beneath another to penetrate the asthenosphere, its movement being accompanied sporadically by earthquakes and volcanic eruptions.

Seismic Waves

As is shown in the diagram on the left, seismic waves are produced by relatively abrupt movements of the Earth's surface as a result of elastic vibration. Such waves, emanating from a zone located inside the crust (hypocentre) are generated by sudden rearrangement of layers of rock. They can be either longitudinal waves (these being the most rapid) or transverse waves (which are slower).

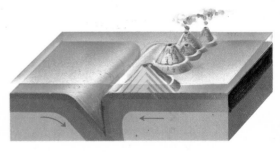

Subduction Zone

A subduction zone is created where one plate is forced down beneath another, thus producing an oceanic trench. The downward-moving plate melts once it reaches the asthenosphere. Volcanic eruptions occur along the margin of the trench.

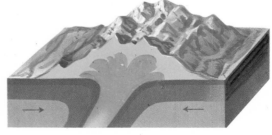

Compression Zones

When two plates carrying continental land-masses come into contact, pronounced folding is produced and this results in the formation of a mountain-chain (orogenesis).

Faults

When two plates undergo a horizontal sliding motion with respect to one another, faults are produced, usually accompanied by intense seismic activity.

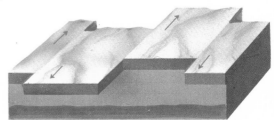

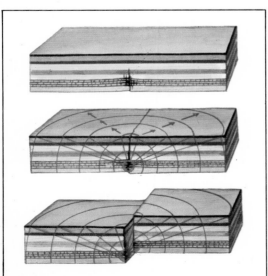

Climate

Climate can be defined as the outcome of interaction between the diverse factors which characterize the Earth's atmospheric layer. Understanding of climatic events requires analysis of the behaviour of the atmosphere over a relatively long time-span. There are three factors to take into consideration: temperature; humidity (and hence rainfall); pressure (which is responsible for the movement of bodies of air, namely winds). In addition to these factors, latitude is also of importance because of its relationship to heat distribution and seasonal variations. On a lesser scale, in terms of microclimate, it is necessary to take account of many other influences which may give rise to variations in the relationships between the basic climatic factors. Altitude, continental influences, exposure, the presence of forests, occurrence of glaciers and other phenomena are all involved. Climate obviously plays a very important part in the modelling of the Earth's landscapes. In fact, any area of the Earth's surface is continually subjected to the action of factors such as ice, heat, rainfall and winds. These alter superficial rocks and bring about a kind of adaptation of the contours of the land and its vegetation to the prevailing type of climate. Man is also subject to the influence of climate, despite the extent to which modern technology permits human beings to tolerate climatic conditions which would otherwise be incompatible with survival. It must also be remembered that the combinations of factors involved have not been stable over time, neither in geological terms nor in historical terms. Geological studies have shown that in past epochs the prevailing climatic conditions were quite different from those found today. For example, in Antarctica geologists have found coal deposits which show that there was once a warm, humid climate in a region now covered in ice. The fact that climate can vary even on an historical scale is demonstrated, among other things, by the retreat of glaciers which has been recorded since the first half of the nineteenth century, resulting from a slight increase in world temperatures.

The Atmosphere
The atmosphere—the gaseous envelope surrounding the Earth—probably developed at the time when the solid outer crust was forming. Some of the main components of the atmosphere (such as nitrogen, carbon dioxide and water vapour) were liberated during the solidification process, whereas oxygen was produced only later by certain early plants. The atmosphere can be regarded as consisting of a series of layers which differ both in physical character and in chemical constitution. The troposphere is the layer in contact with the Earth's surface, varying in thickness from eighteen kilometres (eleven miles) at the equator to six kilometres (four miles) at the poles. This layer is characterized by the presence of air of standard chemical composition familiar to us. Almost all the well-known meteorological phenomena originate in the troposphere, since it is in this layer that vertical movements of air masses and significant quantities of water vapour are to be found. Above the troposphere lies the stratosphere, which has a very low density and is almost completely lacking in water vapour. Above this is the ionosphere, which is characterized by electrically charged particles, and finally there is the exosphere. Beyond that lies the vacuum of interstellar space.

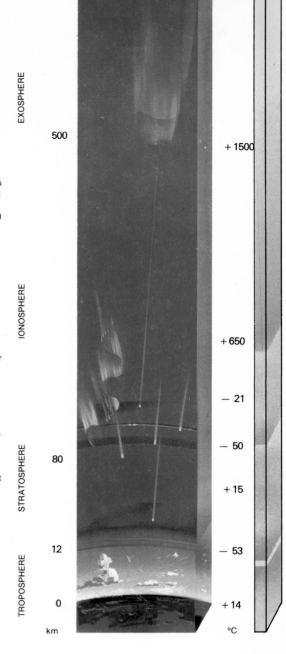

How the Sun Heats the Earth
The illustration (left) clearly shows how the spherical shape of the Earth influences the distribution of heat at the various latitudes. In fact, because of the Earth's shape, the angle between the Sun's rays and a tangent to the Earth's surface becomes more and more acute as one moves from the equator towards the poles. As a result, with increasing latitude the area over which the Sun's thermal energy is dispersed also increases, along with the density of the air which filters this energy. Not only the form of the Earth but also its rotation has an influence on heat distribution, due to the inclination of the Earth's axis and the form of its orbit around the Sun.

Winds
The word "wind" refers to the displacement of air-masses which move from zones of high pressure (anticyclones) to zones of low pressure (cyclones). The pattern of movement is rendered extremely complex by the rotation of the Earth (Coriolis forces), which causes air-masses to deviate to the right in the northern hemisphere and to the left in the southern hemisphere. Differences in pressure can be attributed to marked thermal differences between equatorial and polar regions. Hot air-masses are created near the equator and these move towards the poles, to be replaced by colder air which moves to the equator. Zones of contact between air-masses of different temperature and humidity are characterized by persistent turbulence.

Oceanic Currents
Oceanic currents also constitute an important climatic influence. These currents, which are generated by variations in the density of sea water and by the rotational movement of the Earth, permit a massive transfer of thermal energy from the equator to the poles. The map (left) shows the principal currents in the Atlantic Ocean. It is easy to recognize the northern equatorial current generated by the trade-winds of the Azores region. After passing the Gulf of Mexico, this current swings northwards and exerts a warming influence on European coasts as the well-known Gulf Stream. In the southern hemisphere, the path followed by the equatorial current is different because of interaction with the cold currents of the subpolar regions.

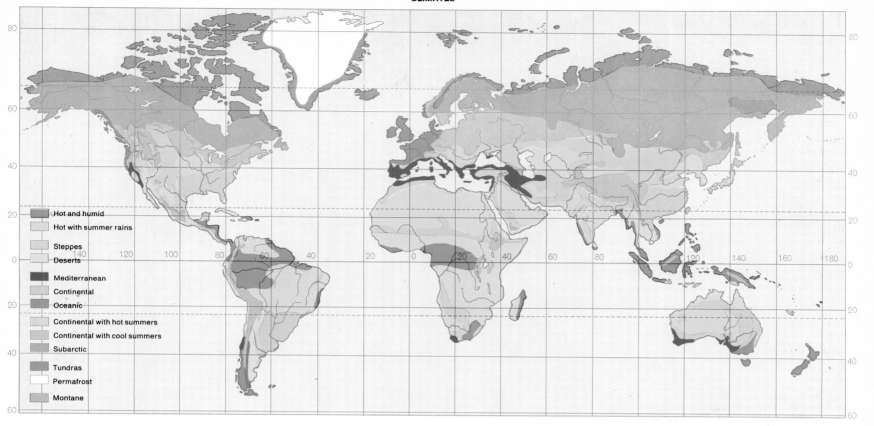

Hot and humid
Hot with summer rains
Steppes
Deserts
Mediterranean
Continental
Oceanic
Continental with hot summers
Continental with cool summers
Subarctic
Tundras
Permafrost
Montane

Climatic Zones

There is an infinite number of possible relationships between the various climatic factors and this obviously creates difficulties for classification. On a global scale, however, three basic types can be recognized: hot climates, temperate climates and polar climates. At a local level, of course, there can be considerable variation in the details of climatic conditions. The most useful classification at present would seem to be that produced by the German climatologist Köppen, as modified by Trewartha.

DISTRIBUTION OF SOLAR ENERGY

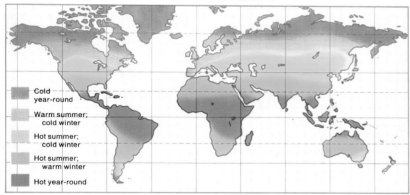

Cold
year-round

Warm summer;
cold winter

Hot summer;
cold winter

Hot summer;
warm winter

Hot year-round

Rain and Solar Energy

As has been seen, the distribution of energy from sunlight and rainfall is a vital factor in determining climate. Regions with a persistent high temperature, notably in the equatorial zone, are characterized by very high rainfall generated by the convergence of warm, humid air-masses. In zones of high pressure (deserts and polar regions), on the other hand, rainfall is very limited.

ANNUAL RAINFALL

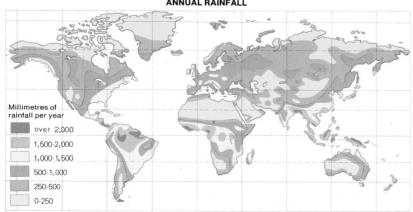

Millimetres of
rainfall per year

over 2,000
1,500-2,000
1,000-1,500
500-1,000
250-500
0-250

The Water Cycle

Rainfall is just one aspect of the water cycle—a complex system through which water is continuously circulated in its diverse physical states, passing from the oceans to dry land and back again. This cycle is fundamental to any form of terrestrial life, which depends upon water as a resource essential to survival. The energy required for this cycle is largely provided by solar heat in interaction with gravitational forces. Thermal energy emanating from the Sun brings about evaporation from the surfaces of the oceans; in other words, it converts water from a liquid to a gaseous state. Part of the water vapour thus formed condenses and falls back into the oceans as rain; but a significant quantity is transported by winds over dry land and is deposited either as rain or as snow. This transported water subsequently gives rise to evaporation from the soil, from lakes, from rivers and from vegetation (plant transpiration). Further, as it flows across the land surface it feeds water-courses, while falls of snow contribute to glaciers. Part of the Earth's water also penetrates the soil and eventually passes into rivers and on into the sea, thus completing the water-cycle.

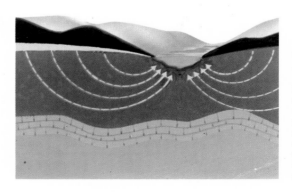

Underground Water

Any water which penetrates into the soil filters through permeable rocks, such as sandstone or limestone, until it reaches impermeable mineral layers. The resulting saturation of permeable rock strata produces an underground reservoir of water which contributes to the river system.

Vegetation

Although we commonly use expressions such as "pine-forest", "chestnut-grove", "beech forest" and "prairie", they are not widely recognized as actual ecological concepts, each defined by a basic set of fundamental constituents. In fact, the analysis of vegetation (that is to say, of the plant assemblages inhabiting different geographical areas) permits us to identify the structure and composition of the diverse habitats which are present on the Earth's surface by virtue of well-defined climatic and biological processes. The modern science of "phytogeography" combines the descriptive analysis of plant species found in association, that is entire plant communities, and the study of the relationships between plants and their general habitats. A phytogeographical study hence involves identification of the plant species in a given area, investigation of the "community structure" (degree of aggregation and isolation of individuals), and subsequent determination of links with climatic and ecological factors of the region concerned. When such studies are conducted systematically on a global scale it is possible to compile phytogeographical maps and to define vast general areas known as "biomes".

Pollination
Plants exhibit three different types of pollination mechanism which ensure fertilization through transport of male gametes (pollen):
1. Wind-borne pollination, as found (for example) in conifers.
2. Water-borne pollination. This is the most ancient in evolutionary terms and is still found, for instance, in mosses and duckweed.
3. Animal pollination. This is a much more sophisticated mechanism confined to flowering plants, which produce nectar to attract an animal carrier (e.g. insect, bird or bat) which will unwittingly transport pollen from one flower to another.

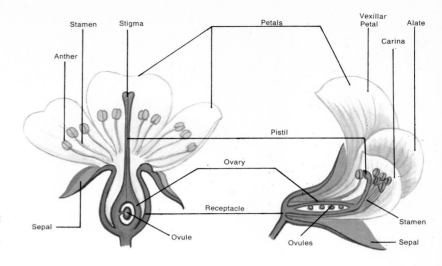

Flowering Plants
When a plant is subjected to the appropriate photoperiod (that is to say, a characteristic relationship between the hours of daylight and hours of darkness) the apical meristems (growing tips) stop producing leaves, thus arresting vegetative growth, and begin a series of transformations which lead to flower development. The leaves themselves are responsible for this transition from vegetative to reproductive growth. In fact, when the leaves have reached their definitive size they also become maximally sensitive to the photoperiod. In response to the appropriate ratio between daylight and darkness, the leaves produce certain substances, including the flower-producing hormone, which reach the apical meristems and induce the production of the characteristic floral structures. The diagram (above) provides two examples of floral organization.
The exceptional capacity of plants to assess the relative duration of periods of daylight and darkness permits all individuals of a species to flower at the same time and thus increases the chances of cross-fertilization.

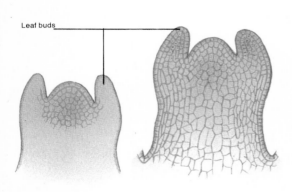

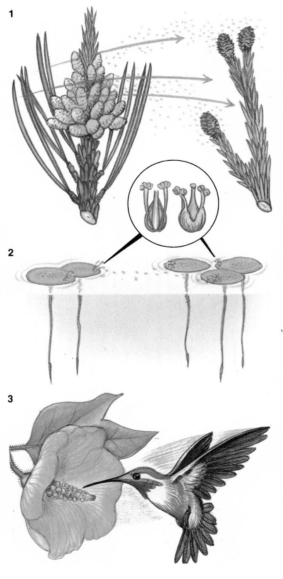

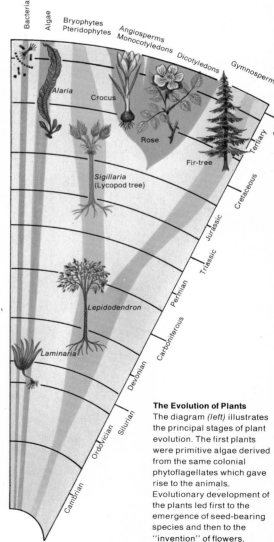

Apical Growth
The growing zone of a bud consists of an apical region where cell-division (and hence an increase in cell number) takes place and of a region of cell-growth (where cell number remains constant) in which the embryonic leaves and the internodal segments of the stem develop until they reach their definitive size. The tissue of the apical meristem (see diagram above) produces new leaves which emerge from the apex in the form of two bulges, each of which subsequently flattens out to form a leaf lamina. Whilst this is happening, the leaf stem grows in length and the vascular elements develop. It should be noted that the entire growth process is governed by plant hormones which respond to external influences such as photoperiod.

The Evolution of Plants
The diagram (left) illustrates the principal stages of plant evolution. The first plants were primitive algae derived from the same colonial phytoflagellates which gave rise to the animals. Evolutionary development of the plants led first to the emergence of seed-bearing species and then to the "invention" of flowers.

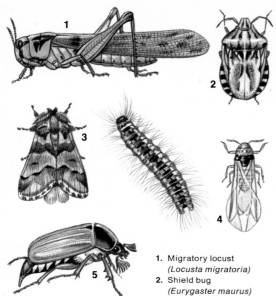

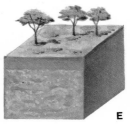

Soils

The term ''soil'' refers to the relatively thin layer, covering the bed-rock, which is generated by the weathering action of atmospheric factors and biological agents. In climatic regions with regular rainfall distributed throughout the year, deep soil-layers are formed, with all the mineral elements evenly dispersed throughout. By contrast, when rainfall is limited in quantity or rare in occurrence during the year, there is only slow modification of rock, and excess water runs across the surface, carrying away the scarce products of weathering. There is also a temperature influence: for any given combination of bed-rock and pattern of rainfall, low temperatures produce a ''podzol'' whereas high temperatures lead to laterite formation. Of course, the chemical composition of the soil is also important in determining the character of any vegetation which becomes established. Typical important chemical components are nitrates, carbonates of calcium and magnesium, and various free ions which determine the level of acidity.

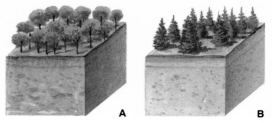

1. Migratory locust
   (*Locusta migratoria*)
2. Shield bug
   (*Eurygaster maurus*)
3. The procession caterpillar and the adult moth
   (*Thaumetopoea processionaea*)
4. Vine-fly
   (*Phylloxera vitifolii*)
5. Cockchafer
   (*Melolontha melolontha*)
6. Adult and larva of Colorado beetle
   (*Chrysomela decemlineata*)
7. Spruce bark beetle
   (*Ips typographus*)

Plant Parasites

In addition to the numerous examples of positive interaction between plants and animals, there are many examples of parasitism, particularly involving insects. It is indispensable for modern agriculture to be able to recognize such plant pests. A number of common ''destroyers'' of vegetation are illustrated *above*. Some insects, such as bugs, suck sap from plants; others deposit their eggs in the meristems of leaves and thus provoke formation of galls; yet others (such as locusts) indiscriminately devour the external parts of plants.

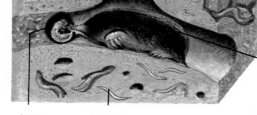

Earthworm

Larva   Nematode   Mole

Aeration of the Soil

The illustration *(above)* demonstrates the importance of various animal species, such as the mole and the earthworm, for turning the soil and hence aerating it. Any terrain which is poor in soil-living animals is always characterized by compact, poorly ventilated soil (e.g. tundra).

A) Deciduous forest
B) Conifer forest (taiga)
C) Tundra
D) Prairie
E) Tropical savannah
F) Desert
G) Typical soil
   1. surface layer
   2. humus
   3. bed-rock

SOIL TYPES

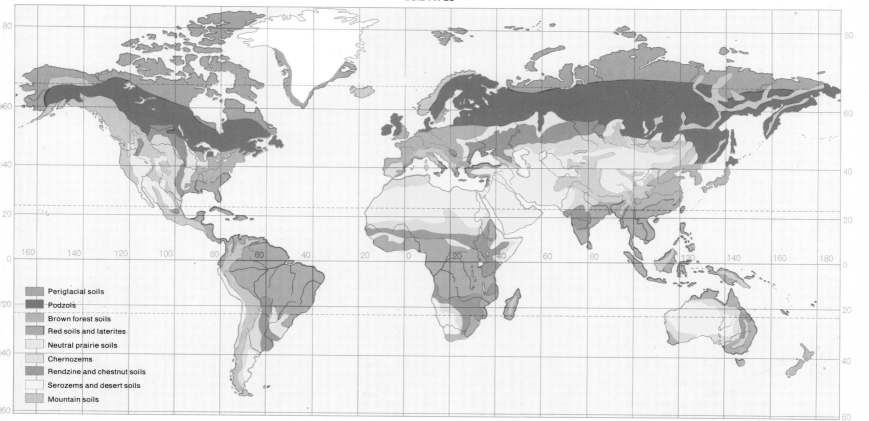

- Periglacial soils
- Podzols
- Brown forest soils
- Red soils and laterites
- Neutral prairie soils
- Chernozems
- Rendzine and chestnut soils
- Serozems and desert soils
- Mountain soils

Evolution

If any given population of organisms is examined it will usually be found that individuals differ from one another at least to some extent, partly because of hereditary differences arising from mutation and genetic recombination and partly because of environmental factors. The structural characteristics of individuals in a population are exposed to the rigours of the environment, and the diversity found within a given species is subject to the filter of natural selection. Those individuals which happen to be best adapted, that is, those which possess the optimal genetic constitution for a given environment, are favoured. Natural selection operates in a multitude of different ways, such as predation, competition for food, climatic factors and differential resistance to parasites. All these different factors combine to select against those individuals which are least suited to the environment and which therefore succumb to an early death or at least suffer some handicap in reproduction. The outcome of natural selection is that the best adapted members of a species survive to reproduce themselves. With every generation there is some readjustment in adaptation, since evolution is, by definition, no more than a statistical process. Further, there are spatial factors such as ecological and geographical influences which are involved in the evolutionary mechanism.

For terrestrial animals, for example, there are ·certain characteristic factors which are involved, notably geographical barriers, such as seas or mountain-chains, which isolate populations from one another and prohibit any genetic exchange between them. In the case of the human species, which probably emerged in Africa before spreading to other continental areas, the origin of diverse ethnic groups depended upon geographical isolation brought about by a number of ecological upheavals which prevented genetic exchange between populations which were originally interconnected. In relatively recent times, rapid increases in transport communication and accompanying increases in the rate of intermarriage between ethnic groups have introduced a new element of genetic change which will probably lead eventually to widespread uniformity of the human race.

Animal Evolution

The present geographical distribution of animals can be considered from a chronological point of view, starting with the origin of life on Earth and proceeding to the present day, by studying the various species which have emerged in succession as a response to environmental change. The chart (right) summarizes the main stages in the evolution of animal life which began with protozoans and led to the array of species living today. According to the most recent interpretations, heterotrophic bacteria (those deriving nourishment from outside) are thought to have given rise to zooflagellates and to have produced a single surviving lineage which terminated in the sponges (Porifera). Another very ancient group of bacteria, with autotrophic (self-nourishing) habits, is thought to have given rise to the phytoflagellates. Colony-living phytoflagellates (of the *Volvox* type) apparently gave rise to two branches, one forming the ancestral stock of the plants and the other leading to the first coelenterates and thus to all other animal species.

The Origin of Life

By recreating the conditions which are believed to have been present on Earth three thousand million years ago, it has proved possible to confirm a basic hypothesis of the origin of life. According to this hypothesis, the action of ultraviolet radiation brought about the transformation of relatively simple molecules—such as methane, ammonia, hydrogen and water vapour—into complex organic compounds which aggregated together to generate structures of ever-increasing complexity, culminating in *coacervates*. Growth and reproduction are the two essential characteristics of living organisms, but we still do not know how simple coacervates developed to produce the first structure definable as a proper "cell".

Animal and Plant Cells

The two diagrams below represent, in schematic form, typical cells *(left)* of an animal, and *(right)* of a plant, showing the fundamental differences between them.
The characteristic structural components of plants are: 1. chloroplasts, containing chlorophyll and the mechanism required for photosynthesis. 2. other pigments, such as carotenoids and xanthophylls, which are also involved in the absorption of radiant energy. 3. a rigid cell wall composed of cellulose. 4. amyloplasts for the accumulation of starch as a carbohydrate reserve.

Animal cells do not possess a rigid cell wall isolating them from the external environment. Instead, they have no more than a cell membrane which in certain cases, as on the outside of the animal, may be hardened to resist mechanical damage. Further, carbohydrate reserves in animal cells are stored in the form of glycogen rather than starch.

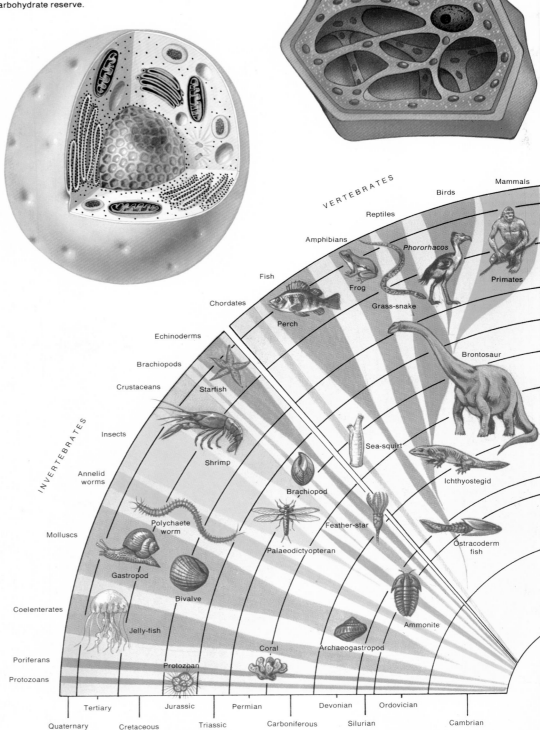

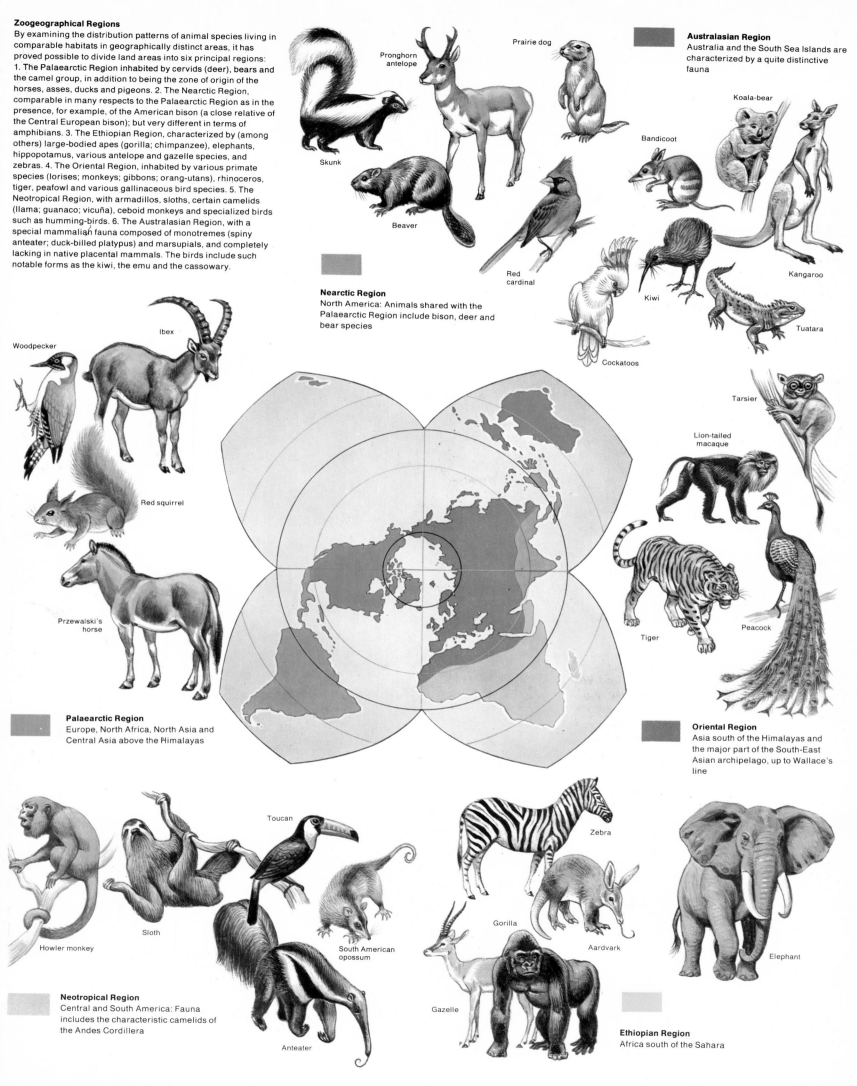

Zoogeographical Regions

By examining the distribution patterns of animal species living in comparable habitats in geographically distinct areas, it has proved possible to divide land areas into six principal regions: 1. The Palaearctic Region inhabited by cervids (deer), bears and the camel group, in addition to being the zone of origin of the horses, asses, ducks and pigeons. 2. The Nearctic Region, comparable in many respects to the Palaearctic Region as in the presence, for example, of the American bison (a close relative of the Central European bison); but very different in terms of amphibians. 3. The Ethiopian Region, characterized by (among others) large-bodied apes (gorilla; chimpanzee), elephants, hippopotamus, various antelope and gazelle species, and zebras. 4. The Oriental Region, inhabited by various primate species (lorises; monkeys; gibbons; orang-utans), rhinoceros, tiger, peafowl and various gallinaceous bird species. 5. The Neotropical Region, with armadillos, sloths, certain camelids (llama; guanaco; vicuña), ceboid monkeys and specialized birds such as humming-birds. 6. The Australasian Region, with a special mammalian fauna composed of monotremes (spiny anteater; duck-billed platypus) and marsupials, and completely lacking in native placental mammals. The birds include such notable forms as the kiwi, the emu and the cassowary.

Australasian Region
Australia and the South Sea Islands are characterized by a quite distinctive fauna

Koala-bear

Bandicoot

Kangaroo

Kiwi

Cockatoos

Tuatara

Skunk

Pronghorn antelope

Prairie dog

Beaver

Red cardinal

Nearctic Region
North America: Animals shared with the Palaearctic Region include bison, deer and bear species

Woodpecker

Ibex

Red squirrel

Przewalski's horse

Palaearctic Region
Europe, North Africa, North Asia and Central Asia above the Himalayas

Tarsier

Lion-tailed macaque

Tiger

Peacock

Oriental Region
Asia south of the Himalayas and the major part of the South-East Asian archipelago, up to Wallace's line

Howler monkey

Sloth

Toucan

South American opossum

Anteater

Zebra

Gazelle

Gorilla

Aardvark

Elephant

Neotropical Region
Central and South America: Fauna includes the characteristic camelids of the Andes Cordillera

Ethiopian Region
Africa south of the Sahara

15

Habit and Habitat

The study of the relationships between a biocenosis (an intimate association of animal and plant species) and the inanimate environment is the province of the modern ecologist. Ecology is a branch of biology which sets out from animal and plant systematics and attempts to provide an overall view of the functioning of each living species in relation to other organisms and to the general habitat in which it lives.

The subject of ecology can be divided into *autecology,* which is concerned with the individual, its way of life and its relations with members of its own and other species within its natural environment, and *synecology,* which is concerned with the analysis of the interactions between all animal and plant populations constituting a biological community in an area.

Autecology includes the study of social organization among the members of a particular species, investigation of any territorial behaviour, analysis of communication systems, and indeed analysis of all the ways in which an organism or group of organisms responds to environmental stimuli. Synecology, on the other hand, includes the study of all those biological mechanisms which contribute to the overall development of a species population, and the investigation of environmental constraints which tend to limit such development.

A knowledge of autecology is an indispensable prerequisite for study of community ecology, since only the study of individuals living in a given habitat will permit us to understand the operation of a food-web or to interpret a diagram portraying the way in which a certain number of unstable communities will tend to result in the same climax (the final stage in a biological progression).

The Ecological Pyramid

Starting from the fact that living things can be divided into autotrophic forms (those which produce organic matter from inorganic resources) and heterotrophic forms (those which consume organic matter), it is possible to construct a simple food-chain which portrays the flow of energy and chemical exchange. On this basis, the following can be distinguished: 1. Producers, the *plants,* which produce organic carbon compounds by means of photosynthesis. 2. Primary consumers, the *herbivores,* which feed directly on plants. 3. An unlimited number of predators, or *carnivores,* which exist by feeding either on herbivores or on other predators. 4. *Degraders* and *mineralizers,* which restore the inorganic material

required by producers for biosynthesis. It is obvious that the number of individuals occupying different levels in this system must be subject to certain constraints. In fact, any plant community has a threshold of exploitation as a food resource above which ecological imbalance will result in destruction of the ecosystem. The relationships involved can be represented by pyramids portraying biomass, energy or numbers of species.

Sea otter
Gyrfalcon
Sea eagle
Skua
Tern
Gull
Oyster-catcher
Sea-perch
Sea bream
Corkwing wrasse
Crab
Squid
Goose Barnacle
Goby
Sea anemone
Gibbula
Limpet
Polychaete worm
Sea-horse
Starfish
Scallop
Sea-urchin
Mussel
Clam
Zooplankton
Tube-worm
Phytoplankton
Green algae
Red algae

Biomes

Even though the climatic and physical conditions of the Earth's surface do not remain constant over time, it is still possible to describe certain communities which tend to be maintained by dynamic replenishment and to develop wherever the appropriate conditions of humidity, temperature, soil chemistry, etc., are available. Such communities are termed "biomes" and they include not only typical plant assemblages but also entire series of animal species whose populations are maintained in balance with the environment. Apart from the obvious marine and freshwater aquatic environments, there are a variety of stable ecological situations which can be found on dry land, ranging from arid desert to luxuriant rain-forest and including a large number of intermediate biomes reflecting the variety of available climatic conditions.

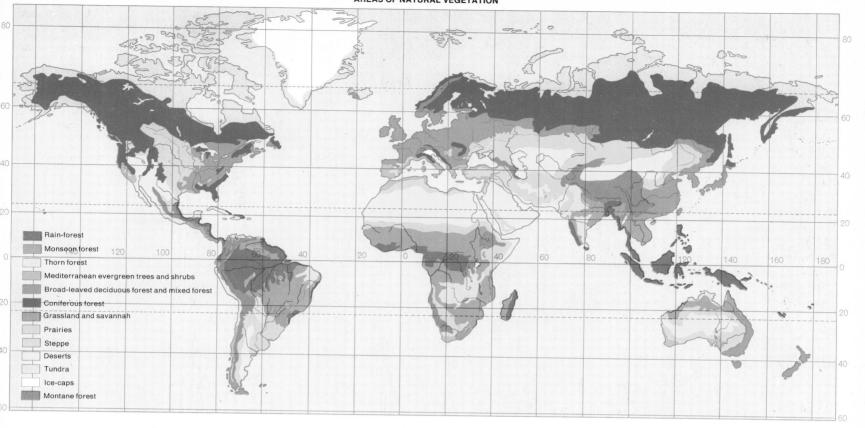

Rain-forest
Monsoon forest
Thorn forest
Mediterranean evergreen trees and shrubs
Broad-leaved deciduous forest and mixed forest
Coniferous forest
Grassland and savannah
Prairies
Steppe
Deserts
Tundra
Ice-caps
Montane forest

Ectoparasites

Among the many examples of negative interactions between species, is the interesting case involving man and the ectoparasitic tapeworm. The term ectoparasitism is used here because this ribbon-like worm lives in a cavity—the intestine—which is actually in communication with the external environment. The intermediate host of the tapeworm is the pig; man becomes infected by eating contaminated pork.

Endoparasites

Endoparasites, which live inside parts of the body which do not communicate directly with the external environment, include the well-known example of *Plasmodium,* the protozoan responsible for malaria. Human beings are actually intermediate hosts for this endoparasite, since it reproduces asexually within the human body. The *Anopheles* mosquito is the definitive host in which the malarial parasite exhibits sexual reproduction.

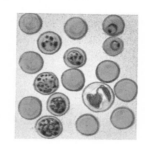

Symbiosis

An example of perfect symbiosis is provided by termites and hyper-mastiginous zooflagellates. In fact, these two species are unable to exist without one another. The termite is only able to obtain its food from wood because the protozoan, which inhabits the intestine of the termite, breaks down the cellulose. Similarly, the zooflagellates are unable to lead a free-living existence if they are not encysted. Whenever moulting occurs, the termite discards its old skin and at the same time loses the lining of its stomach enclosing the zooflagellates.

Subsequently, the termite would be unable to digest cellulose if it did not rapidly take steps to re-establish its precious culture of protozoans by ingesting encysted zooflagellates.

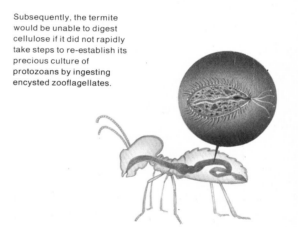

SPITZBERGEN

ARCTIC OCEAN

NOVAYA ZEMLYA

TAIMYR

NORWEGIAN SEA

LAPLAND

KOLA PENINSULA

TUNDRA

TIMAN RANGE

CENTRAL SIBERIAN PLATEAU

ICELAND

S C A N D I N A V I A

FINLAND

Lake Onega

Lake Ladoga

North Dvina

U R A L S

Ob

Yenisei

WEST SIBERIAN PLAIN

Yenisei

Angara

Lena

NORTH SEA

BALTIC SEA

Lake Baikal

BRITISH ISLES

NORTH EUROPEAN PLAIN

CENTRAL RUSSIAN UPLANDS

VOLGA UPLANDS

Volga

Dnieper

Ural

Ob

Irtysh

SAYAN

YABLONOWY RANGE

A T L A N T I C   O C E A N

Loire

Rhine

Danube

Volga

KAZAKH UPLANDS

MONGOLIA

Mont Blanc

ALPS

CARPATHIANS

Lake Balkhash

GOBI DESERT

PYRENEES

BALKAN

BLACK SEA

CAUCASUS

Aral Sea

CASPIAN SEA

TIEN SHAN

GREAT KHINGAN

IBERIAN PENINSULA

MEDITERRANEAN SEA

ANATOLIA

ARMENIA

TURKMENISTAN

TAKLA MAKAN

K U N L U N

Hwang-Ho

SZECHWAN

Euphrates

SYRIAN DESERT

IRAN

H I M A L A Y A

A F R I C A

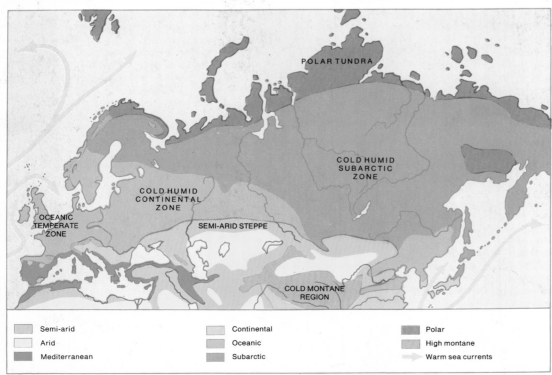

POLAR TUNDRA

COLD HUMID SUBARCTIC ZONE

COLD HUMID CONTINENTAL ZONE

OCEANIC TEMPERATE ZONE

SEMI-ARID STEPPE

COLD MONTANE REGION

Semi-arid	Continental	Polar
Arid	Oceanic	High montane
Mediterranean	Subarctic	Warm sea currents

EURASIA

Europe is often considered to be a self-contained entity, but in fact it is a peninsula of Asia. In the natural sciences, Eurasia is regarded as a genuine geographic unit. Eurasia is an immense continent and everything seems to be adapted on a suitable scale. The largest forest in the world, the taiga, is found in this region, extending from Scandinavia to the Pacific coastline. A unique and impressive mountain-chain passes from one side of Eurasia to the other, beginning with the Pyrenees in the west and passing through the Alps, the Carpathian Mountains, the Caucasus and the mountains of Armenia and Iran to end with the gigantic Himalayan chain, which contains the highest peaks in the world. As far as biogeography is concerned, Eurasia is divided into two distinct zones, the Palaearctic Region (consisting of Europe and northern Asia) and the Oriental Region (consisting of South and

South-East Asia). The Palaearctic Region is characterized by low temperature to temperate conditions, with subtropical climates occurring only in the Mediterranean area. Eurasia contains numerous "biomes" in addition to the Arctic tundra and the taiga. There is a succession of deserts spanning the area from Arabia to China; vast steppe zones occupy the central part of Eurasia; broad-leaved deciduous forests occur in Europe and in Manchuria; and there is a special plant association surrounding the Mediterranean Sea, known as the maquis. Among the most representative animals are the well-known European bird species and a variety of mammals such as red deer, fallow deer, members of the goat and cow groups, wild asses, camels, antelopes and large carnivores such as the brown bear, the wolf, the Siberian tiger and the leopard.

The Extreme North

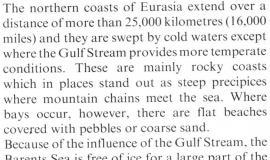

The northern coasts of Eurasia extend over a distance of more than 25,000 kilometres (16,000 miles) and they are swept by cold waters except where the Gulf Stream provides more temperate conditions. These are mainly rocky coasts which in places stand out as steep precipices where mountain chains meet the sea. Where bays occur, however, there are flat beaches covered with pebbles or coarse sand.

Because of the influence of the Gulf Stream, the Barents Sea is free of ice for a large part of the year and is therefore inhabited by a rich and varied animal community. Mussels, barnacles, sponges and sea-squirts find refuge in the rock crevices, while numerous species of crabs, shrimps, sea-urchins and starfish live among the algae on the sea-bed. The open waters are well stocked with different fish species, the most common being cod, herring and sole. These fish provide food for a large number of aquatic mammals such as cetaceans (whales and killer whales), pinnipeds (seals and sea-lions) and sea-otters, as well as for a host of birds including loons, guillemots, puffins and gulls which nest along the coast. The beneficial effect of the Gulf Stream gradually decreases as one moves eastwards; beyond Novaya Zemlya the sea is covered by ice virtually the whole year round. The animals inhabiting this region of the coast are correspondingly few in number. In the Bering Sea, the waters become more temperate once again thanks to the warming effect of the North Pacific Current. South of the Bering Straits, the sea contains thriving populations of molluscs, echinoderms and crustaceans.

The Arctic tundra forms a broad swathe extending up to the northern coastline and ranging from Scandinavia to the Kamtchatka peninsula. The tundra is characterized by very low annual temperatures and by the freezing of the soil throughout most of the year. During the brief summer, the soil unfreezes just at its surface to form a network of marshes. The vegetation comprises only lichens, mosses and robust herbaceous plants such as saxifrages, clubmosses and sedges.

The Nordcapp
The northern limit of Norway, lying above the 70th parallel.

The Reindeer's Food
Various reindeer subspecies inhabit the northern regions of Eurasia *(right)*. These animals feed on mosses and lichens, travelling continually over long distances to find new feeding grounds.

Iceland
The valleys of Iceland are carpeted in heathland, meadows and pastureland *(below right)*.

Cortical layers

Hyphae (Medullary layer of the fungus)

Unicellular algae

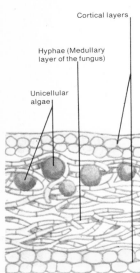

A Favourable Association
Lichens colonize bare rock surfaces and are able to withstand quite harsh climatic conditions. They are produced by a symbiotic association between a fungus and an alga, within which the chlorophyll-bearing alga furnishes the fungus with sugar in exchange for water and mineral salts. In this way, both the fungus and the alga are able to colonize sites where they would be unable to survive in isolation.

Extremely Brief Summers
In the northern tundra, which is characterized by low temperatures and icy winds with the result that the soil is almost always frozen, the period available for plant growth is very limited. For most of the year, plant roots are unable to absorb water and they are accordingly subject to arid conditions quite like those found in deserts. All the plants are stunted in size, and annuals are virtually non-existent. Indeed, large expanses of soil remain completely bare. Only in sheltered places can briefly flowering campions, saxifrages and dwarf willows *(right)* be found, all of which are also typical inhabitants of alpine regions.

20

Animal Life in the Tundra

The richness of animal life in the tundra zone varies markedly from season to season. In summer, when the sun is out round the clock, the lakes, rivers and ponds thaw and a multitude of insects (such as mosquitoes) and other invertebrates can be found among the vegetation. At this time, many bird species—geese, ducks, swans, waders, buntings, thrushes and larks—and a number of large herbivores, such as the reindeer and the elk, migrate northwards, attracted by the abundant supply of food. When winter approaches, however, the tundra is abandoned by most of its animal visitors. The birds fly off to the tropics, while the reindeer and elk take refuge in the taiga zone—an immense forest area characterized by boreal (northern) conifers. Arctic hares and hazel-grouse also move southwards, followed by their natural predators, the foxes, wolves and wolverines. Only a few animal species remain in the tundra for the winter. The insects over-winter as eggs or larvae, whereas small rodents dig burrows beneath the snow and stock them with food reserves. Among the larger mammals, only the great polar bear remains to hibernate in a den hollowed out of the frozen snow and thus survive the harshest winter months, waking only occasionally to seek food.

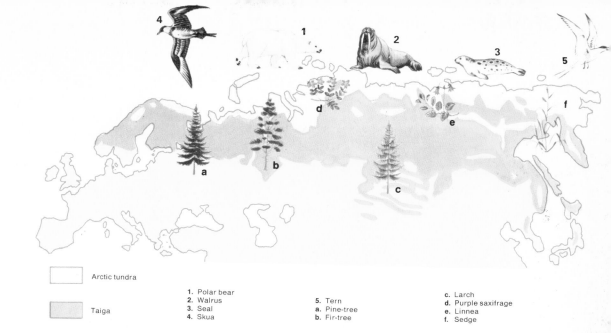

Arctic tundra

Taiga

1. Polar bear
2. Walrus
3. Seal
4. Skua
5. Tern
a. Pine-tree
b. Fir-tree
c. Larch
d. Purple saxifrage
e. Linnea
f. Sedge

The Arctic Pinnipeds

Pinnipeds (seals and sea-lions) are nimble swimmers, but they move around in awkward fashion on land, albeit with an occasional surprising burst of speed. These mammals are perfectly adapted for life in the sea; they return to the coast only for breeding. Their diet consists predominantly of fish and crustaceans, while they are themselves preyed upon by killer-whales and polar bears. The latter lie in wait for them by holes in the ice, through which the prey must emerge in order to breathe. *Left:* A group of sea-lions.

The Coastline of Spitzbergen *(above)*

The islands of Spitzbergen, lying between 70° and 80° latitude north, are characterized by an Arctic climate tempered by the Gulf Stream. The mountains and glaciers together produce an alpine landscape. The peaks are not particularly high, but the rock faces are precipitous and owe their distinctive shape to the successive glaciations of the Quaternary period. There is one peculiar feature of this archipelago which is characteristic of the Arctic landscape: the glaciers, which are otherwise similar to those found in more southerly mountain-chains, sometimes descend almost to the sea-line.

An Arctic Tundra Landscape

The U-shape of the valley *(below)* clearly shows the important influence exerted by glaciation on the shaping of this landscape. Until about 8,000 years ago, these regions were covered with a thick layer of ice which was dispersed by subsequent climatic changes. At the foot of the rocky escarpments lies a thick covering of detritus composed of rock fragments, split off by the action of the ice, which progressively built up to form vast deposits.

Sea-birds

During the short-lived summer, the rocky coasts and islets become the home of millions of sea-birds which return to dry land to make their nests. Guillemots, loons, gulls and puffins form colonies containing thousands of individuals, with their nests packed closely together. This gregarious nesting behaviour is an effective defence mechanism. Any predator which attempts to rob a nest is confronted by a barrage of beaks, wings and screeches which will disorient the invader and often put it to flight. *Below:* A colony of gannets; large sea-birds which nest on islands just off the coast.

The Boreal Forest

The greatest forest in the world, the taiga, consists of boreal (northern) conifers and covers a total area of more than twelve million square kilometres (4.5 million square miles), ranging from Scandinavia to the coasts of the Pacific Ocean. The forest is interrupted here and there by heathland, enormous marshes and a network of small lakes. Despite the apparent poverty of plant species, the taiga provides a home for a great number of animals, including hordes of insects and other invertebrates, since it provides food throughout the year: seeds, buds, shoots and bark. Even during the long, harsh winters, there are many animals which remain active beneath the carpet of snow. This is, for example, the case with several insectivores such as the mole and the tiny shrew. Some birds—swans, woodpeckers and owls—nest in tree-hollows, while others, such as wood-grouse and hazel-grouse, build their nests among roots on the ground. Woodpeckers are agile hunters of insects. Cross-bills, with pincer-like beaks, make a speciality of winkling out pine-seeds.

A number of species store food for the winter. Rats and squirrels stock seeds in their nests, the red squirrel hiding its seeds in tree cavities. The Siberian jay buries pine-cones in the ground. Only the brown bear and the badger become lethargic in winter. The typical carnivores include the lynx (which preys principally on hares) and the wolverine, which will also eat carrion. But the commonest carnivores are the marten (a major predator of squirrels), the weasel and the stoat. In Siberia, the sable is also a common predator. Raptors include the golden eagle, the buzzard, the goshawk and the Ural owl. The marshes are inhabited by swans, sandpipers, ducks, wagtails and yellowhammers. The commonest raptor is the hen harrier. There is also a large variety of fish species: carp, pike, perch and salmonids.

Porsangen: One of the Northernmost Fjords of Norway
Fjords are one of the most characteristic features of the landscape at high latitudes. These long, narrow inlets of the sea can sometimes penetrate dozens of kilometres inland and they owe their origin to submersion of glacial valleys. The sides of fjords can sometimes form stark precipices and they are peculiar in that the depth of the water remains virtually constant from one end to the other, though there is almost always a rise in the rocky bed at the opening to the sea. It is also common to find that fjords have numerous side-branches. All these characteristics can be attributed to the glacial origin of fjords; similar characteristics are to be found in alpine valley systems. One of the longest fjords, located just to the north of Bergen, is the branching Sogne fjord, which extends 160 kilometres (100 miles) inland and has an almost constant width of five kilometres (three miles) from end to end. Fjords are common not only along the Norwegian coast, but also in Sweden, Greenland, Labrador and Chile.

The Origin of Fjords
The illustrations (above and right) show three phases in the development of the fjords which now indent the Scandinavian coastline. Before the ice spread, the landscape was characterized by rolling hills and V-shaped river valleys. During the Ice Ages, enormous glaciers carved sharp precipices in the mountains and the valleys became U-shaped. After the retreat of the glaciers, rising sea-levels led to invasion of the valleys by sea-water to produce the fjords.

The Bjornoff Fjord
A platform of floating logs on the Bjornoff Fjord in Sweden (above).

The Baltic Sea during the Ice-Ages
During the Quaternary period, an ice-sheet of enormous dimension covered the entire Baltic region and neighbouring areas. As the ice melted and retreated, the sea-level rose and sea-water penetrated the valleys originally created by the glaciers.

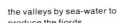

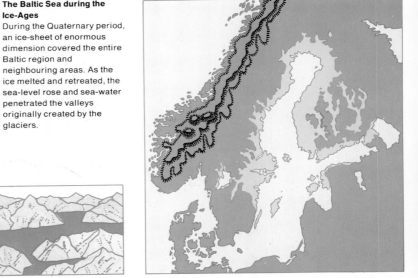

Lake Baikal
In south-west Siberia, a region characterized by a continental climate, lies Lake Baikal, which plays a major part in shaping the biological character of the entire neighbouring region. This huge mass of water in fact has a thermo-regulatory effect and renders the climate more temperate. The extremely clean lake-water provides a home for a large variety of animal species, including sponges, tube-worms, gastropod molluscs, crustaceans, twenty-three fish species and a single mammal —the Baikal seal. Because of the high density of crustacean predators, the fish which live in the lake have developed special breeding behaviour. In order to protect their eggs, they either swim up river-courses before spawning or lay their eggs beneath stones and remain alongside to protect them. *Below:* The Ural owl and a clump of edelweiss.

The Brown Bear
The brown bear *(right)* once distributed throughout Europe, is now extremely rare and the largest remaining population is found in Russia. Brown bears are solitary inhabitants of the forest, feeding on berries and other plant foods. Only occasionally does this bear species feed on meat, obtained by hunting elk, reindeer or ailing red deer.

The Herbivores
The herbivorous mammals inhabiting the taiga can be quite large. During the coldest months of the year, elk, red deer, reindeer and European bison take refuge in the boreal forest. Their natural predators, the wolves, rarely penetrate into the forest, preferring to remain in open areas. *Above:* A Siberian ibex.

Long Winters
The conifer forest is almost as cold as the tundra and the plant species exhibit a number of special structural features which permit them to survive long periods of water shortage when the soil is frozen. Despite this, the boreal forest "biome" is characterized by the presence of very tall trees: Siberian pine, Scots pine, Norway pine, larch. With the exception of the larch (which loses its leaves during the coldest part of the year), the conifers are evergreen trees. Their roots are relatively shallow, but they spread widely through the upper soil layer. They are well protected against the freezing conditions thanks to the reduced level of transpiration guaranteed by the waxy coating of the leaves and by the deep location of their stomata (respiratory pores). In places where availability of water is not such a serious problem, broad-leaves trees can be found mingled with the conifers: birches, alders and willows. The undergrowth is quite sparse in places and typically includes junipers, bilberries, mosses and lichens. In marshy areas, cotton-sedge is a typical plant inhabitant.

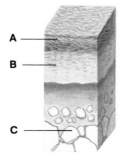

Soil Structure
In an environment where the snow layer persists for seven or eight months a year and where the average temperatures during the coldest period drop below −30 °C, the soil is characterized by an acid humus. The intense cold inhibits bacterial degradation, with the result that the conifer needles remain in the upper layer of the soil for a long time while they are slowly broken down.

A) Acid humus
B) Zone of accumulation
C) Bed-rock

Mixed and Broad-leaved Forests

Below the northern taiga belt, Europe was once covered, from the Atlantic coast to the Urals, by rich forests containing broad-leaved, deciduous trees interspersed with marshes and clearings with herbaceous ground cover. Nowadays, a large part of that original forest has disappeared, to be replaced by cultivated land and urban areas. In this region, the climate is quite humid with rainy summers and cold winters. Beyond the Urals stretch the immense, arid high plateaux of Asia; as a result, the deciduous forest does not reappear until the Far East is reached, where the proximity of the ocean lends a more humid character to the climate once again. The great European mountain-chains—the Pyrenees, the Alps and the Carpathian Mountains—are included in the deciduous tree zone, but their high altitude also favours the development of conifers. The undergrowth of deciduous forests is substantial and provides food for a wide variety of animals; indeed, animal life flourishes at every level from the ground up to the canopy. Insects

The Soil of Mixed Forest
The diagram *(right)* shows the structure of the soil typically found in temperate zones subject to heavy rainfall. It is characterized by intense bacterial activity, by the presence of an alkaline humus or mull (A) and by an organic layer rich in minerals (B, C). The humus is progressively mixed with the mineral fraction, since the decomposers continually turn the soil.

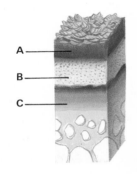

The Broad-leaved Forest
The most distinctive feature of this forest, as compared to the boreal forest, is its average annual temperature between 5 °C and 10 °C, with the highest value of 22 °C during the hottest months and a minimum of − 10 °C in the coldest months. Rainfall is quite heavy and favours the development of maple, elm, lime-tree and walnut, in addition to encouraging a rich undergrowth, woody creepers such as ivy and herbaceous plants such as the hop. This plant community is associated with oceanic climates and is widespread in Europe, but less common in the East, where it is replaced by steppe until the Pacific Ocean makes its influence felt in the Far East.

Distribution of Conifers and Broad-leaved Trees in Europe

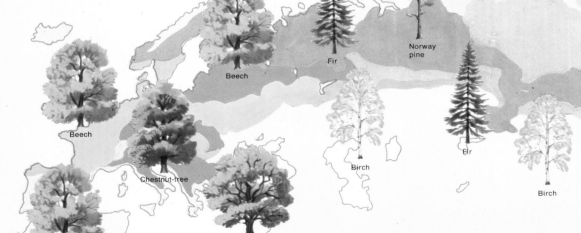

Norway pine

Fir

Beech

Beech

Beech

Chestnut-tree

Oak

Birch

Fir

Birch

Larch

Fir

Maple

Oak

Beech

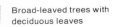
Broad-leaved trees with deciduous leaves

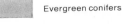
Broad-leaved deciduous trees and evergreen conifers

Evergreen conifers

Deciduous conifers

nd other invertebrates are particularly numerous and their activities play an essential ole in the life-cycle of the forest. The soil is nhabited by various worm species which turn nd aerate the soil. They also carry down to he subsoil layer plant fragments which are ransformed, through the action of decomosers (fungi and bacteria), into the inorganic natter required by plants. Fallen tree-trunks, narsh vegetation and carrion are also decomosed by numerous invertebrates to yield impler substances which are attacked in turn by fungi and bacteria. The remaining animal population of this "biome" is constituted by a variety of birds, rodents, insectivores and natural predators (weasels, stone-martens, polecats and badgers). Large predators, such as bears, wolves and lynx, which were once common in these forests, are now virtually absent. In the deciduous forest of the Far East, which is otherwise comparable to the European forest in terms of its animal inhabitants, red deer and roe deer are hunted by the powerful Siberian tiger. Hares, pheasants and wood-grouse provide the common prey for panthers and lynx, while the weasels prey upon squirrels and rats.

The Undergrowth
The predominance of deciduous trees over the conifers is associated with a greater variety of plant species in the undergrowth, such as the primrose *(above)*, the cuckoo-pint and the carnation *(right)*.

The Animals of the Forest
Because of the abundant food and the opportunities for shelter provided by broad-leaved forests, they are inhabited by a number of snakes and lizards, by rodents (field-mice and voles), by insectivores (moles and hedgehogs) and by bats. The densest areas of the forest are also inhabited by red deer and wild boar. In summer, birds are very common in such forests. Tree-creepers, nightingales and blackbirds hunt there for insects. Jays, bullfinches, tits, goldfinches and wood-pigeons feed on berries and seeds. The red fox *(right)* is probably the commonest predator, feeding primarily on small rodents, which are also preyed upon by stone-martens, polecats, weasels and badgers. The raptors include goshawks, buzzards *(below)* and sparrow-hawks, all of which hunt for reptiles, birds and amphibians. Nocturnal birds of prey are also common. The largest is the eagle owl, which preys upon hares, rabbits, mustelids (weasels, etc.), squirrels, rats and hedgehogs.

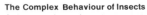

The Complex Behaviour of Insects
Quite a number of insects are pests: caterpillars feed voraciously on leaves; cockchafer larvae feed on roots, and so on. The damage they cause is limited by their predators—carnivorous insects and spiders. *Above left:* An adult cockchafer. *Above right:* A bumble-bee. *Below left:* A house spider. *Below right:* A cabbage white butterfly.

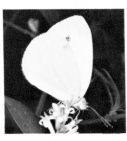

Migratory Birds
When winter approaches, many birds take wing for the tropics. Storks *(right)* take their leave of Central Europe and head southwards by two main routes. Some of them fly over the Iberian peninsula to reach the area around Lake Chad in Africa, while others cross the Balkans to reach the same destination by following the course of the Nile. Yet other storks pass the winter in Arabia or in Southern Africa.

25

The Forests of the Far East

The Animals of Szechwan

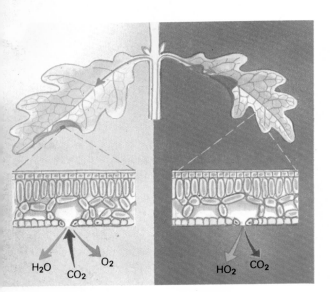

The fauna of this region is characterized by unusual animal species, including a number of strange-looking bovids and a collection of primitive insectivores. Some of the insectivores resemble moles and live in underground burrow systems, while others look like shrews but are active swimmers and live close to rivers. The best-known animal of this region is the giant panda *(above)*, which feeds exclusively on bamboo shoots. The lesser or red panda *(right)* is also a typical inhabitant of the bamboo forest.

Szechwan

The slopes of the high plateaux of south-west China are covered in broad-leaved forest, which contains an increasing proportion of conifers at higher altitudes. At still higher altitudes, there is a vast zone where rhododendrons and bamboos grow *(above)*. This is the Szechwan region, which is one of the richest areas for animal species in the cold, temperate or tropical regions of Asia, since it is located at the junction between the two great biogeographical regions of the continent, the Palaearctic region and the Oriental region. Variation in climatic conditions is the main characteristic feature of Szechwan, since the altitude varies from 1,000 to 7,500 metres (3,000 to 25,000 feet). While the upper reaches are covered with snow and tundra, the valleys enjoy a subtropical climate. The mountains, which run in a north-south direction, present a barrier to the humid winds coming from the west. As a result, the sky is always cloudy in Szechwan and it rains frequently. As a rule, there are no more than thirty days of clear sky in the whole year and this region is known to the Chinese as "the land of clouds".

The Fuji-San Volcano

This, the most famous of the Japanese volcanoes, has been inactive since 1700. It bears witness to the geological instability of the Japanese archipelago and to its recent origin. The Fuji-San volcano rose up during the mountain-building phase of the Tertiary period, which itself led to the emergence of the Japanese island chain. The volcano is venerated as a sacred mountain and frequent pilgrimages are made there.

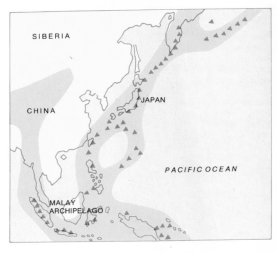

"The Ring of Fire" of the Pacific *(above)*
This is the name given to a series of volcanoes, many of which are active, which fringes the western margin of the Pacific Ocean. This volcanic zone extends from the Kuril Islands in the north of Japan down to New Guinea.

Photosynthesis *(left)*

Plant life is fundamentally dependent upon the process of photosynthesis, which can be divided into two phases:
1. *The photochemical response,* by means of which the indispensable energy of sunlight is captured by the chlorophyll and used for the chemical cleavage of water. The energy liberated by this cleavage is stored in molecules with a great reducing power (NAD-PH and ATP). Oxygen is liberated as a by-product.
2. *The dark reaction,* which takes place within the matrix of the chloroplast and involves a reaction between NAD-PH and carbon dioxide to produce, through a complex series of stages, carbohydrates which are stored by the plant.

Leaf Structure *(right)*

In higher plants it is the leaf which acts as the specialized centre for photosynthesis. As shown in the diagram, the leaf contains the following layers (proceeding from top to bottom): A protective layer of closely packed cells (epidermis) followed by one or more layers forming the palisade tissue, composed of elongated, vertically oriented cells. Beneath this is the spongy parenchyma, which plays an important part in gaseous exchange with the outside world. Between these two inner layers of the leaf run the veins, which transport nutrients.

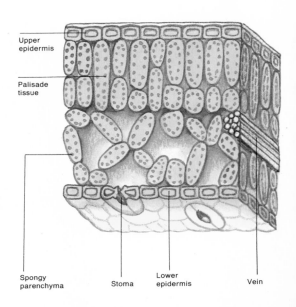

Upper epidermis

Palisade tissue

Spongy parenchyma

Stoma

Lower epidermis

Vein

H_2O O_2 CO_2 HO_2 CO_2

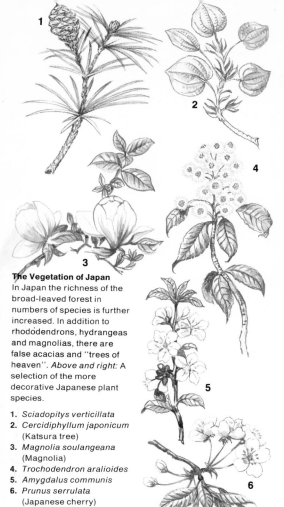

The Vegetation of Japan

In Japan the richness of the broad-leaved forest in numbers of species is further increased. In addition to rhododendrons, hydrangeas and magnolias, there are false acacias and "trees of heaven". *Above and right:* A selection of the more decorative Japanese plant species.

1. *Sciadopitys verticillata*
2. *Cercidiphyllum japonicum* (Katsura tree)
3. *Magnolia soulangeana* (Magnolia)
4. *Trochodendron aralioides*
5. *Amygdalus communis*
6. *Prunus serrulata* (Japanese cherry)

Plant Nutrition

In any higher plant two main structural components can be recognized: the root, which anchors the plant in the soil and functions to absorb water and mineral salts, and the shoot. The diagram *(below)* demonstrates the plant's basic system of nutrient supply. Transpiration through the leaf pores (stomata) requires a supply of water and salts, which are transported from the roots along the xylem (woody tissue). The roots and the trunk, on the other hand, cannot carry out photosynthesis and require a supply of sugars produced by the leaves, which is transported along the phloem (soft tissue of the stem).

Xylem (woody vessels transporting crude sap)

Phloem (sieve element vessels transporting elaborated sap)

Absorption of water and salts by the root hairs.

The Animals of Japan

The geographical position of Japan is associated with great variety in climatic conditions and plant species, but animal life is not so varied as on the neighbouring mainland. Overall, the common species of the mixed forest of Manchuria are predominant in Japan. On the northernmost islands can be found Siberian tigers, lynx, wolves, wolverines, reindeer, foxes, sable, weasels, otters and brown bears. The brushwood is inhabited by local wild boar species. The sika is a medium-sized deer species confined to Japan, with a dark coat which is dappled with white in summer. The most characteristic animal is the Japanese macaque *(above)*. The islands of Japan support a rich bird fauna, including jays, tits, goldfinches, thrushes, woodpeckers and sparrows in the broad-leaved forests and crossbills, hawfinches and francolins in the conifer forests. In the tropical areas of Japan, pheasants and ant-thrushes are common and in marshy areas there are a number of typical members of the heron family such as the great egret *(right)*. The mountain streams have a rich population of fish, crustaceans and molluscs, and there is also the Japanese giant salamander which can grow to more than 1.5 metres (five feet) in length.

The Crater of the Zaô-San Volcano

The Zaô-San Volcano, located in northern Japan on the island of Honshu, contains a small lake within its crater, showing that it has been inactive in recent times. Japan has been aptly described as "the land of volcanoes and earthquakes". Within historical times, dozens of active volcanoes have been recorded and each year the Japanese archipelago is shaken by innumerable seismic waves. Japan lies on a particularly unstable part of the Earth's crust, being located at the junction of two of the tectonic plates which constitute the lithosphere (the Eurasian plate and the Pacific plate). The Pacific plate is very slowly moving down beneath the Eurasian plate and, in the course of geological time, this movement has given rise to a number of mountain-building events as well as accounting for the emergence of the Japanese archipelago itself. The downward movement of the Pacific plate has also been associated with the development of extremely deep oceanic trenches, such as the Japan trench and the Kuril trench.

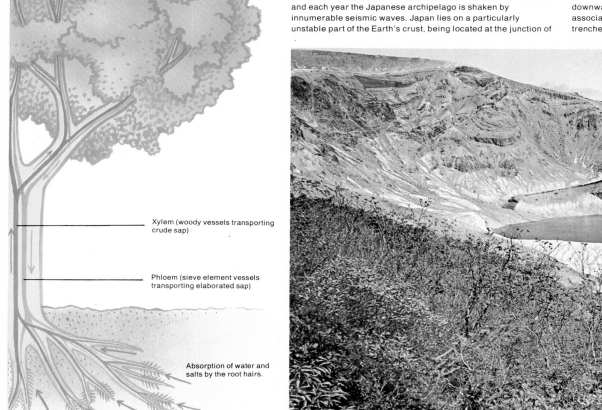

Mountains of Eurasia

Although they occupy only a small proportion of the Earth's surface, mountains exert a strong influence on climate by determining prevailing wind-directions and thus patterns of rainfall. The great Eurasian mountain-chains are oriented along an east-west axis and thus create clear-cut climatic divisions along their slopes, which are enhanced by differences in latitude. Moving up the slopes, broad-leaved forests give way to mixed forests and finally to conifer forests, and each forest zone is inhabited by characteristic animal species. Above the tree limit, which varies in altitude according to latitude (600 metres or 2,000 feet, in Scandinavia; 1,500–2,400 metres, or 5,000–8,000 feet, in the Alps), there is a shrub zone which is followed by a herbaceous zone which itself eventually gives way to a zone where only mosses and lichens grow.

At high altitudes, there is permanent snow cover with conditions of life which are similar in many respects to those of polar regions, but the atmosphere is more rarefied. Any animals which live at this altitude must be able to cope with a number of problems: scarcity of food, harsh climate, shortage of oxygen and isolation. Many plant and animal species which occur at high altitude are, in fact, subarctic species and their presence bears witness to the recent Ice Ages. Taking the principal Eurasian mountain-chains in turn, the first is the ancient Scandinavian chain in northern Europe, where the most distinctive pioneer tree is the birch and where the animal fauna is particularly well represented by birds such as the Norwegian gyrfalcon, the golden eagle and the snowy owl. Mammals are relatively rare; the best-known is undoubtedly the lemming, a small rodent which carries out periodic, spectacular migrations. The lemming is preyed upon by raptors and by a number of carnivores, including the stoat, the polecat, the lynx and the wolverine.

The Pyrenees and the Spanish Sierras
In stark contrast with the fertile valleys and marshy areas, the sierras and high plateaux of southern Spain are generally quite arid because of the erosion brought about by the action of the wind and rain, often exacerbated by human activity and the depredations of domestic goats. This habitat is occupied by a variety of small rodent species, a number of small passerines, swallows, swifts, crows and alpine choughs. Higher up, there are griffon vultures, Egyptian vultures, kestrels, golden eagles and buzzards. The large mammals include isolated groups of ibex and the occasional wolf. Trout are common in the rivers and streams. The Pyrenees (above) are similarly quite desolate in appearance, though there is more plant cover than in the sierras and they have a thriving animal life despite the relatively high altitude, averaging more than 3,000 metres (10,000 feet). The entire region is rich in mountain torrents, lakes, woods, pastures and alpine precipices. Animal inhabitants include chamois, bears and a wide variety of bird species, such as thrushes, wood-grouse, woodcock, ptarmigan and vultures.

The Eurasian Mountain-chains (below)
The map shows the major orographic lines of Eurasia. It can be clearly seen that the geological relief is essentially based on a single chain, the Alpine-Himalayan system (also known as the circum-mediterranean system). This chain is a gigantic series of mountain massifs which begins in the west with the peaks of the Iberian peninsula (or, more exactly, with the African Atlas mountain-chain) and proceeds through the Pyrenees, the Alps, the Carpathian Mountains, the Caucasus and the mountains of Armenia and Iran to reach the great Himalayas, which include the highest mountain peaks in the world. All these mountain regions are relatively new in geological terms. They were formed during the Cenozoic (Tertiary) era as a result of the slow northward movement of the African and Indian tectonic plates. This movement gave rise to compression and folding in the geosyncline (a large depression in the Earth's crust) located in the area of the modern Mediterranean region.

The Structure of Mountain-chains
The diagram (below) illustrates two fundamental types of mountain-chains: those produced by folding and superposition, and those produced by fracturing and dislocation. Tectonic forces can produce either the folding of rock layers (folded structure with synclines and anticlines) or a system of fractures. In the latter case, when the fractured blocks are elevated or depressed, the term fault is used.

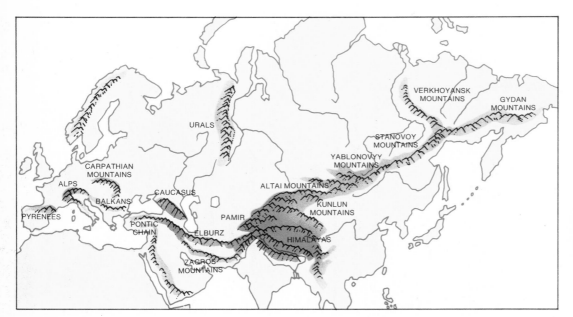

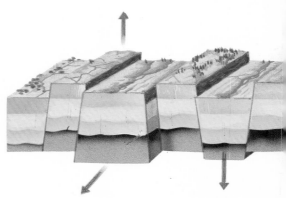

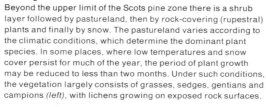

The Large Herbivores of the Mountains Red deer *(top and far right)*, ibex *(above left)* and chamois *(above right)* spend the summer feeding on pastures and in woods high up on the mountains. In the coldest winter months, however, they descend to the valleys.

The Vegetation

Beyond the upper limit of the Scots pine zone there is a shrub layer followed by pastureland, then by rock-covering (rupestral) plants and finally by snow. The pastureland varies according to the climatic conditions, which determine the dominant plant species. In some places, where low temperatures and snow cover persist for much of the year, the period of plant growth may be reduced to less than two months. Under such conditions, the vegetation largely consists of grasses, sedges, gentians and campions *(left)*, with lichens growing on exposed rock surfaces.

Animal Life

There is a much greater density of animals on mountain slopes covered with conifers and broad-leaved trees, which provide both food and protection throughout the year. Above the tree limit, living conditions are rendered difficult by isolation, by the scarcity of food and by rarefication of the atmosphere. The air is poor in oxygen, a gas which is indispensable for animals since they need it to produce energy through breakdown of the food they ingest.

The rarefied atmosphere contains very little water and it permits greater penetration by ultraviolet rays, which can be harmful to living organisms. As a result, any animals which live at these high altitudes have developed special adaptations, such as a thicker pelage to provide protection against the cold and against ultraviolet radiation. It is also common to find a greater degree of development of the heart and lungs than in comparable species living at low altitudes, and there is a greater concentration of red blood corpuscles in the circulating blood as an adaptation to increase the uptake of oxygen by the haemoglobin. *Right:* A marten. *Above:* Its usual prey, a red squirrel.

The Caucasian Mountain-chain *(right)*

The Terskol valley leads to the foot of Mount Elbrus, which has the highest peak of the entire Caucasian chain (5,633 metres, or 18,476 feet). The landscape of the Caucasus is of the alpine type, with sheer escarpments and glaciers above the tree limit. This mountain-chain was formed at the same time as the Alps, as part of the great Alpine-Himalayan folding event which occurred during the Tertiary era.

The high plateaux of Scotland, an extension of the Scandinavian mountain-chain, show similar characteristics and possess a nordic fauna and flora. These plateaux are almost completely lacking in trees and are typically covered in alpine heather which provides shelter for wood-grouse, partridges, hares, martens and golden eagles.

The Alps are one of the youngest mountain-chains of Europe. Although the wooded slopes constitute a distinctive zone, the animal inhabitants are species which also occur elsewhere, since the many mountain passes provide easy routes for dispersal. Nevertheless, there are a number of species which are typically found far away in northern Europe and which also occur in the Alps as vestiges of the Ice Ages, such as wood-grouse, ptarmigan, dotterel, hazel-grouse and certain insects. The pastureland is inhabited by marmots, ibex and chamois, while in the alpine lakes and water-courses live trout, char, grayling and white-fish. The Apennines, a geological extension of the Alps, have woods containing oak, ash and hornbeam. In the highest reaches of the Abruzzi (in Central Italy) wolves, chamois and a few remaining bears can be found. The Carpathian mountains, which were originally an eastern extension of the Alps, are inhabited by numerous small rodents and by a variety of larger mammals such as red deer, roe-deer, chamois, wild boar, bears, wolves and lynx. The water-courses of this region are well-stocked with trout and with Danube salmon. The vast forests of the Caucasus contain red deer, roe-deer, wild boar, bears, polecats, wildcats and lynx, while the pastureland is inhabited by chamois, ibex, distinctive rodents such as the souslik (a kind of ground-squirrel), particular butterflies such as the mountain clouded yellow, and numerous raptors including the spotted eagle. The occasional leopard may also be found there.

The Pokkara Valley *(above)*
This valley in Nepal lies in the foothills of some of the highest mountains of the Himalayas, such as Annapurna, Manaslu and Dhaulagiri. There are fourteen mountain peaks which exceed 8,000 metres (26,000 feet) in height and they are all to be found in the Himalayas or in the Karakorum—the two Asiatic mountain-chains which together form the greatest mountain region on Earth. Together, these chains cover a range of 2,500 kilometres (1,600 miles) in length and an approximate width of 300 kilometres (190 miles).

The Formation of the Himalayan Mountain-Chain *(right)*
As with most of the other great mountain-chains, the formation of the Himalayas resulted from the collision of two of the Earth's tectonic plates. In the case of the Himalayas, this collision involved the Eurasian and Indian plates. The latter was originally united with Africa and Australia some two hundred million years ago and then began to drift slowly northwards until it collided with the Eurasian continental land-mass to produce a series of parallel folds.

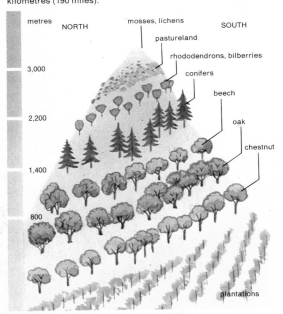

Stratification of Vegetation according to Altitude
The stratification of vegetation with altitude is a direct function of humidity, temperature and the duration of snow cover. This relationship is illustrated on a grand scale by mountain-chains which are oriented in an east-west direction, such as the Himalayas and the Alps. These chains have one slope facing south—which is warmer, drier and has less persistent snow cover; and the other slope facing north, with resultant lower temperatures, greater humidity and greater duration of snow cover. The diagram *(left)* shows how the zones inhabited by broad-leaved, deciduous trees (which require somewhat higher temperatures to prosper) are more closely packed on the southern slopes. The same effect is repeated going uphill through the conifer zone and through the pastureland to reach the "pioneer" vegetation level.

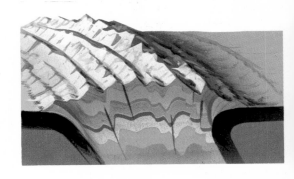

The Himalayas

The gigantic mountain-chain formed by the Hindu Kush and the Himalayas stretches from the Mediterranean to the Bering Sea and clearly separates the cool, temperate zones of Asia from the tropical regions. This enormous rampart has an average altitude in excess of 3,000 metres (10,000 feet) and through a succession of higher chains and elevated plateaux it rises almost as high as 9,000 metres (30,000 feet) in the areas of its greatest peaks. Whereas the high plateau zone has a uniform appearance and an essentially arid character, the peripheral mountain-ridges exhibit a great variety of animal- and plant-life. In fact, there is a notable difference between those slopes which are exposed to sea winds, which generate high rainfall, and the enclosed slopes, which are arid. The most widely distributed and best adapted animals in the high reaches of the Himalayas are the various well-known cow-like (bovine) and goat-like (caprine) mammals. *(See illustration below.)*

| Chiru 5,400–6,000 m (18,000–20,000 ft) | Yak 4,200–6,000 m (14,000–20,000 ft) | Markhor 1,000–2,000 m (3,000–6,500 ft) | Takin 3,500–5,000 m (11,500–16,500 ft) | Tahr 2,000 m (6,500 ft) | Marco Polo sheep 2,700–5,500 m (9,000–18,000 ft) | Bharal 3,600–4,500 m (12,000–15,000 ft) | Tibetan gazelle 2,700–3,600 m (9,000–12,000 ft) |

The Hoofed Mammals of the High Plateaux
The extensive open areas on the arid high plateau of Tibet provide an ideal habitat for numerous hoofed mammal (ungulate) species, since natural predators can be spotted from some distance away and thus are avoided in good time. The illustration shows just some of the Himalayan ungulate species. The photograph *(above right)* shows three yaks, which have a shoulder height of 1.65 metres (5 ft 5 in). In the background can be seen Annapurna, a mountain 8,078 metres (26,650 feet) high.

On the Mountains of Asia
The forests lining the mountain-chain stretching from Elburz to the Hindu Kush are very rich in animal-life, including many species which are common in Europe. The deciduous woodlands provide a home for various deer species, wild boar, wolves, tigers, leopards and a variety of bird species. The conifer forests are inhabited by squirrels, woodpeckers and slender-billed nutcrackers.
Reptiles are also common up to an altitude of 2,000 metres (6,600 feet). There are various lizard species and the noteworthy Himalayan Halys pit viper, a member of the rattlesnake subfamily related to the moccasin of the New World. The mountain-chains of Karakorum and the Himalayas constitute a southern barrier for many species of the temperate region and, at the same time, a northern barrier for many tropical species. Predators include the Asiatic black bear *(below)*, the snow leopard and numerous raptors such as the lammergeyer. The Siberian mountains are also inhabited by certain species which are common to the tundra (reindeer, lemmings, snow-hares) and a number of other typical inhabitants of the taiga (lynx, wolverines, red foxes).

Mountain insects
1. Apollo butterfly
2. Mayfly
3. Mayfly larva
4. Stonefly
5. Caddis-fly larva
6. Springtails
7. Athomyiid flies
8. Rove-beetle
9. Carabid beetle
10. Ladybirds
11. Earwig

Insects of the High Mountain Regions
Insect species which live at high altitudes depend heavily upon the winds sweeping up from the plains to obtain their food. These winds carry up the mountain-slopes pollen and seeds, as well as other insects and spiders which die immediately they come into contact with the freezing air.

Life in Cold Deserts

From the Euphrates to the slopes of the Greater Khingan range, Asia is marked by a wide diagonal band of temperate and cold deserts, characterized by very hot summers and harsh winters. Despite the arid conditions, there is a flourishing animal population, though this is greatly influenced by the short period of plant-growth and the scarcity of watering-places. Large herbivores, for example, are obliged to keep moving continually. In order to reduce water-loss to a minimum, rodents, small carnivores and birds typically remain hidden during the daytime and become active at night. In winter, however, this behaviour is reversed: small mammals and birds only emerge from their retreats during the middle of the day. The land tortoise and ground-squirrels of the genus *Citellus*, which feed on green plant parts, are only able to feed from March to May; for the rest of the year they remain torpid. The commonest animals are lizards and geckos, which provide an important link in the desert food-chain. These reptiles feed on insects and spiders, while they themselves are preyed upon by mammalian carnivores, raptors and snakes. The deserts contain a variety of venomous snake species (vipers and cobras) and sand boas of the genus *Eryx*. Insects are even more abundant and diversified. Large predatory ants and harvesting ants keep constantly on the move, while termites attack practically everything that is not actually made of metal; they combine to produce a veritable plague. Turning of the soil, which in other regions is accomplished by earthworms, is carried out by a plant louse of the genus *Hemilepistes*. A hectare contains more than a million of these lice; they construct underground tunnel-systems which permit aeration of the soil and penetration by rainwater.

The Steppe
The heart of Eurasia is constituted by an uninterrupted broad band of grassland which stretches from Iran to Mongolia. This is the steppe zone, characterized by long, icy winters and very hot summers. The steppe is a veritable ocean of grass beaten by violent winds and changing in colour from season to season. At the beginning of the summer, the grass is pale green, but by August it has become straw-coloured. To the north, the steppe gives way to the taiga, whereas to the south it borders the desert.

The Arid Lands

The commonest mammals of the steppe are rodents, whose active burrowing habits produce a beneficial turning action in the soil layer. In addition, rodents carry into their burrows organic material, such as grass, which is rapidly converted into humus. The humus layer is accordingly exceptionally thick in such soils. Nevertheless, proliferation of the rodents is limited by the activities of a large number of predator species: wolves, foxes, weasels, stoats, badgers, wildcats, a very few leopards and a variety of raptors such as the golden eagle, the buzzard, the kestrel and the powerful eagle owl.

Large herbivorous mammals also contribute to the ecological equilibrium of the steppe. Continually on the move in search of new grazing areas, these herbivores hammer the soil with their hooves, thus driving seeds below the surface, in addition to cropping the grass and limiting its growth. Finally, their droppings provide valuable manure for the soil. In previous centuries, the vast steppe zones provided a home for enormous roving herds of saiga, wild horses, wild asses, gazelles, antelopes and camels. Nowadays, these original populations have been drastically reduced by hunting and some are close to extinction, though roe-deer, saiga and Asiatic moufflon are still relatively common. Wild boar are also common in the very dense vegetation bordering the large rivers—the Dnieper, the Don, the Volga and the Ural. A variety of insectivorous birds— little bustards, demoiselle cranes, shrikes and starlings—are able to feed on the hordes of insects ranging from bees to locusts. There are also many different seed-eating bird species: partridges, quail, Pallas's sandgrouse, larks and finches.

The Desert of Afghanistan
The photographs (above left and left) show two views of the Afghan high plateau, which is characterized by vast rocky areas with cold climatic conditions, especially in winter.

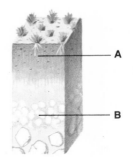

Steppe Soil
The essential characteristics of the steppe soil are the presence of a uniform humus layer at the surface (**A** above) and the existence beneath this of a layer of calcium carbonate (**B**). In this soil, organic matter is produced not through slow degradation by fungi but through rapid combustion.

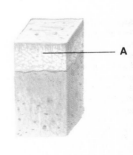

Desert Soil
In the desert, a combination of strong winds, high temperatures and low rainfall effectively precludes formation of a deep soil. Desert substrates are poor in organic matter and the soil horizons are hardened owing to the presence of limestone, gypsum or clay (**A** above). Salts, particularly chlorides, are present in high concentrations as a result of continuous evaporation.

Steppe Animals

Since there is no shelter available above ground, many steppe-living rodents, insectivores and small carnivores are obliged to dig burrows to escape from predators and the harsh climatic conditions. The same applies to reptiles and amphibians and even some birds—desert wheatears, shelduck and ruddy shelduck—take refuge in abandoned burrows. Mole-rats, small rodents which have become adapted like moles, never emerge above ground and feed upon roots and corms. In order to escape the snow in winter, the large herbivores carry out great migrations over vast distances, while family groups of wild boar retreat into bushy areas and depend upon the layer of subcutaneous fat which they have accumulated during the summer. Wolves, which are able to run across the snow, prey upon saiga and roe-deer. *Above:* Demoiselle cranes. *Right:* A pratincole.

The Bactrian Camel

Wild asses, such as the onager *(right)* and camels, are able to survive for long periods without drinking, by extracting water from the plants they ingest.

The Cold Deserts *(below)*

The cold deserts together form a broad band running from east to west, including the Gobi Desert, Tibet and Turkestan. They are very cold regions with a continental climate and annual rainfall of less than 300 millimetres (12 inches) a year. This aridity of the climate is due to the fact that the summer monsoons are blocked on the other side of the Himalayas.

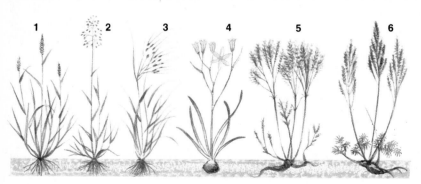

Steppe Plants

The principal plant species found in the steppe are grasses and flowering plants with corms, which contain a rich reserve of food substances. Brambles, tamarisk bushes and acacias can also be found here and there. *Above:* The commonest plants of the steppe: **1.** *Poa crestata* **2.** *Poa bulbosa* (bulbous blue grass) **3.** *Stipa capillata* **4.** *Gagea bulbifera* (gagea) **5.** *Artemisia maritima* (white wormwood) **6.** *Artemisia pauciflora* (black wormwood)

The Vegetation

In these arid regions, where rocky outcrops are common and where sand is continually moved about by wind action, plant growth is severely restricted. The vegetation largely consists of annual species which depend upon the very short humid periods in the year to complete their life-cycles.

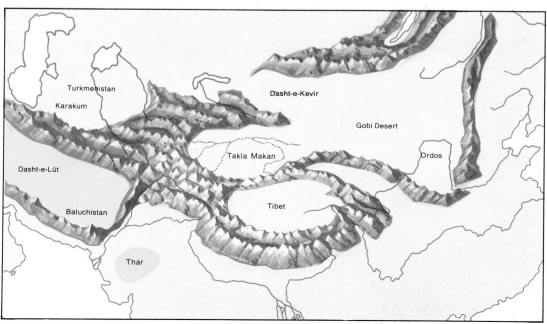

Mediterranean Region

The Mediterranean region has a subtropical climate and the vegetation is extremely varied, containing both woodland species and typical desert plants. The ancient Mediterranean forest, which was characterized by established trees such as oaks, conifers and olive-trees, is now considerably fragmented because of the massive changes brought about by human activity over the centuries. Of the various plant associations to be found, the commonest is the maquis, which is composed of small trees, bushes and shrubs. The same type of vegetation is also found in other zones of the world where comparable climatic conditions prevail, notably in South Africa, in California, in Chile and in southern Australia. The animal-life of the maquis is still incompletely documented, but it is known that human modification of the vegetation has also had a deleterious effect on the animals. Most of the large carnivores have disappeared; all that is left is a small population of leopards in the Atlas region of Morocco, and a few surviving bears in the mountains of north-west Spain and in the central part of the Apennines. The wildcat is rather better represented. The Mediterranean fauna also includes a monkey species in North Africa, the Barbary macaque, a small population of which has been artificially established in Gibraltar. In addition to a large number of insect species, including many different beetles, the invertebrates are represented by a variety of scorpions, millipedes, spiders, snails and slugs. The rich population of invertebrates attracts a corresponding number of predator species, such as various lizards and a great number of birds, which are the most commonly noticed inhabitants of the maquis.

The Flora
The flora of the Mediterranean region is characterized by the presence of evergreen tree species such as the holm-oak, the mastic-tree, the cane-apple and the wild olive-tree. On siliceous substrates, the holm-oak is replaced by the cork-oak and on flat terrain the Aleppo, Corsican and maritime pines are found.

Where the soil is poor, kermes-oak can also be found. The Mediterranean maquis is now dominated by an assemblage of bushes, myrtle, Spanish juniper, heather and mastic-trees. Further away from the coast towards drier areas, the maquis slowly gives way to open patches of terrain and eventually grades into the garrigue—a plant formation characterized by a thin soil layer and the exposure here and there of the underlying rock. This formation is typified by the presence of *Genista* (a relative of broom), rosemary, thyme and lavender. The animal community which inhabits this association has not yet been adequately studied, but the invertebrates of course include beetles, scorpions, millipedes, snails and slugs, while the characteristic mammals are the wild rabbit, the wildcat and the porcupine.

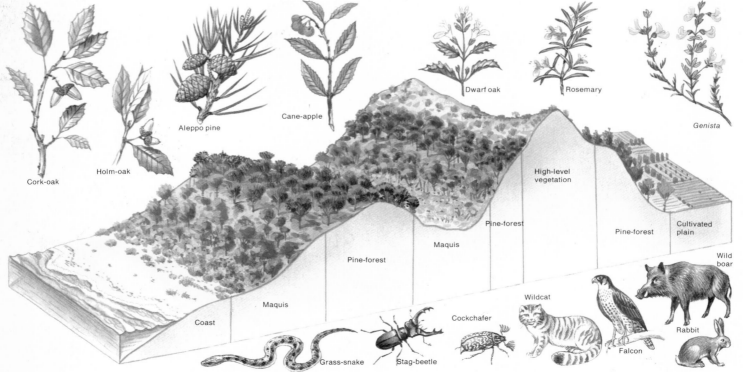

Cork-oak · Holm-oak · Aleppo pine · Cane-apple · Dwarf oak · Rosemary · *Genista*

Coast · Maquis · Pine-forest · Maquis · Pine-forest · High-level vegetation · Pine-forest · Cultivated plain

Grass-snake · Stag-beetle · Cockchafer · Wildcat · Falcon · Rabbit · Wild boar

The Olive-Tree

The evergreen Mediterranean maquis, which is found in regions with a temperate but rather dry climate, is above all characterized by large groves of olive-trees. Indeed, the distribution of the olive-tree exactly traces the limits of the maquis. The photograph *above* shows a plantation of cultivated olive-trees, which grow extremely slowly and may live for centuries. The inflorescences of the olive-tree are distinctive in that they are grouped in clusters. Because its value has long been appreciated, the cultivated olive-tree now occupies a considerably larger geographical range than its wild relative.

The Grape-Vine

It would scarcely be suspected from a glance at the grape-vine *(above)* that in the natural state it behaves like a woody creeper and uses other trees as supports by wrapping its tendrils around them. It is for this reason that grape-vines are normally propagated by cuttings in vineyards and seed is used only to produce new varieties.

Tobacco

Five centuries ago, the tobacco plant was known only to certain Indian tribes of the New World as a material for smoking; now it is used throughout the world. Initially, tobacco was used in Europe only for medicinal purposes; the smoking of tobacco for pleasure did not begin to spread until the end of the seventeenth century. Cultivation of tobacco requires a warm, rather humid climate. Because of their high nicotine content, tobacco extracts are also used as insecticides.

Wild Sheep

Herds of moufflon live on the rocky slopes of Sardinia and Corsica, often in areas which are inaccessible to human beings. Moufflon, like other ruminants, have a particularly keen sense of smell and can detect potential enemies at a great distance.

A Destructive Parasite

The female olive-fly lays her eggs in the fruits of the olive-tree by means of her piercing ovipositor. The larvae which develop then feed on the fruit-pulp of the olive. *Below:* The reproductive cycle of this destructive parasite, which occurs throughout the area inhabited by olive-trees.

The Distribution of the Olive-Tree

The map *above* shows the distribution of the olive-tree in the Mediterranean basin.

Adult

Egg deposited in an olive

Larva

Pupa, with the adult fly just hatching

The Tireless Singer

Although it is universally known for its uninterrupted singing, the cicada *(right)* is actually quite difficult to observe. Because of its grey-brown colour and its habit of remaining almost completely still, it easily escapes detection.

The Aleppo Pine

This pine-tree *(below left)* is a characteristic inhabitant of flat areas around the Mediterranean, though its present distribution has been considerably influenced by human intervention.

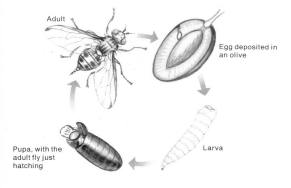

Nocturnal Activity

During the daytime, the wild rabbit *(right)* remains hidden in its burrow, which has a number of tunnels and a chamber well below ground. The rabbit comes out to forage only at night.

InlandWaters

The rivers, lakes and ponds of Eurasia provide a great variety of ecological niches and accordingly contain an extremely diverse fauna. The commonest animals are invertebrates, which are able to exploit all the opportunities offered by the watery environment. They live and breed among the gravel and the sand, between the fronds of plants and even in open water. Invertebrates play an essential role in maintaining the ecological balance of freshwater habitats. Some species are herbivorous, others are carnivorous and yet others feed on decomposing organic matter. The most widely distributed invertebrates are the single-celled protozoans. Rotifers, nematodes and a large number of crustacean species are also present. The abundant and varied food resources provided by lakes and rivers are also exploited by numerous fish species, many of which have developed special adaptations to this end. Some species have mouths modified for water-filtration: they feed off small particles of organic matter, while other fish feed on plants; still others have become active hunters. By contrast, there are very few mammals associated with freshwater habitats. As a rule, mammals tend to construct their retreats on dry land and only visit the water to feed. Usually, they have webbed paws, and a thick coat of fur to protect them from the cold. In Eurasia, the principal mammals found in inland waters are the water-vole, the desman, the otter and the beaver.

The Amphibians
There is a considerable variety of amphibian species, all of which rely on water for egg-laying. Since they are unable to maintain a constant body temperature, they are most common in warm temperate zones. Typical European species are the common frog, the tree-frog *(right)* and the salamander. *Below:* Two dragon-flies preparing to mate, and a grass-snake.

Fast-flowing Streams
The currents in fast-flowing streams present a considerable problem for animals and only certain invertebrates and good swimmers, such as the trout, are able to live in them. The water in such streams is cold and rich in oxygen, but plant-life (with the exception of algae) is very limited.

River-Courses

Rivers are one of the major agents involved in the progressive modification of landscape. In the upper reaches of a river, steep inclines lend considerable energy to the water. As a result, there is marked erosion of valleys through the scouring away of bedrock material which is transported downstream. However, as the land becomes flatter, a river loses a great deal of its energy, regardless of how much it is fed by affluents, and it develops a meandering course where deposition predominates. During such deposition, the river selectively gives up the material it has accumulated. Large fragments of rock are deposited first, then pebbles and, finally, where the river flows out into the sea, silt which leads to delta-formation.

Glaciers

River-Sources

Alluvial meander

Delta mouth

UPPER REACHES

Deposition cones

MIDDLE REACHES

Affluents

Incised meander

LOWER REACHES

Ox-bow lake

Rivers
The upper reaches of rivers have fast-flowing water, but silt deposition does occur along stretches where vegetation is present. In such places, a variety of animal species can be found. There are a number of fish, such as the grayling and the minnow, and there are many insect larvae which provide the primary food for fish. As the river progressively descends to the plain, the current becomes slower and silt deposits become more extensive. Floating vegetation and substrate-living plants become much more prevalent, providing both food and shelter for many fish species, such as the pike *(above right)*, the perch *(centre right)* and the carp *(below right)*. Close to the mouth of the river, the water becomes brackish and the only fish species to be found there are those which are able to live in both fresh water and sea-water.

Marshland

Marshes provide an environment which is intermediate between a lake and a meadow, and they play an important role in containing and regulating river floodwaters. They also provide a home for certain species, such as herons, avocets and wild duck, which are unable to build their nests elsewhere. The most striking feature of this habitat is its vegetation, which forms distinct zones in concentric circles. In a sequence passing from the bank out into the marsh, there may be a fringe of water-loving trees (willows and poplars), then a zone of plants which are more extensively adapted to life in the water (peat-mosses and marsh marigold), and subsequently plants which live more-or-less submerged (bulrush, reeds, water-lilies). Finally, in the centre of the marsh itself, there are genuine floating plants whose roots have no contact with the substrate, such as duckweed, and which often have leaves that differ in form according to the depth of the water, as in the case of the arrowhead. In an environment of such diversity, the animal-life is also greatly diversified. Numerous herbivores and carnivores are involved in the complex food-chain of marshland. In addition to otters, frogs and newts, there are a great many insects: dragonflies, mosquitoes, water-beetles and water-boatmen are typical. The food-cycle is closed by biodegraders, which return inorganic material to the primary producers—the plants. Particularly active in this respect are a host of bacteria and the annelid worm *Tubifex*, which lives entirely buried in the silt and generates a large quantity of mud.

The Camargue and Las Marismas

West of Marseilles, the river Rhône forms a large marshy zone of subtropical type, the Camargue. This region is inhabited by many migratory bird species, the most representative being the pink flamingo. The mammals include the famous wild horses *(above)*, black cattle living in a semi-wild state and wild boar. Rodents abound and they attract a variety of predators, such as owls, foxes and weasels. Another natural zone which has a rich fauna is the region between Seville and Cadiz in Spain, lying just inland from the delta of Guadalquivir. The region is known as Las Marismas (literally: "the marshes"). Although it is completely flooded in winter, in summer it is transformed into a vast heathland area in which red deer, wild boar and fallow deer are found along with the rare Spanish lynx. The bird-life is particularly rich, including geese, ducks, avocets, sandpipers and various lark species. Raptors are also common, such as the imperial eagle, buzzard, kite, short-toed eagle and griffon vulture. In Romania, not far from the mouth of the Danube, there is yet another vast area of marshland which is inhabited by an enormous number of aquatic and shore-living birds.

The Marshlands of Europe *(right)*

Marshes are very common in the north and in Siberia, where they were formed when the ice retreated, but they become progressively rarer to the south. In southern regions, marshes are only found close to deltas of the great rivers: Guadalquivir, Rhône, Po, Danube and Volga.

The Kingfisher

A well-known example of a bird which makes its living from rivers is the kingfisher, which is distributed throughout Europe, Asia, Africa and New Guinea. The kingfisher feeds by perching on an overhanging branch and diving down to catch any fish that it spots. The prey is always swallowed head first.

Animals
1. Perch
2. Carp
3. Pond-snail
4. Frog
5. Sticklebacks
6. Pike
7. Grass-snake
8. Kingfisher
9. Mallard
10. Water-beetle
11. Water Scorpion
12. Dragonfly
13. Otter
14. Short-toed eagle
15. Egret

Plants
a. Arrowhead
b. Water-lily
c. Hornwort
d. Pondweed
e. Duckweed
f. Bulrush
g. Reed
h. Horsetail

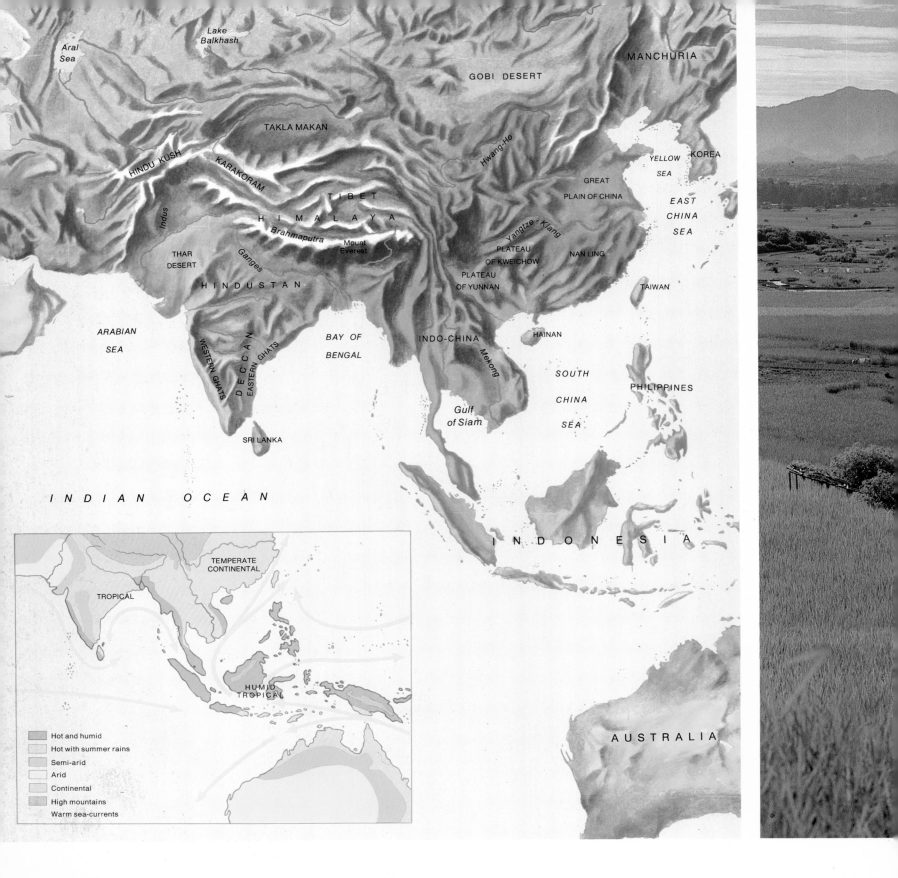

TROPICAL ASIA

Cut off from the rest of the continent by the gigantic Himalayan mountain-chain, southern and South-East Asia form the Oriental region, including India, Sri Lanka, the Indochina peninsula, Malaysia and the Philippines. The climate of this zoogeographical area is warm and humid. India provides a meeting-ground between the west and the east, and this is reflected in the plant and animal species found there. For example, the high central Deccan plateau has a temperate climate and its forests, which have now been greatly reduced by human intervention, consist of broad-leaved deciduous trees. By contrast, long stretches of the coast of India are characterized by mangrove forests. In South-East Asia, the rainforest no longer forms a continuous belt as it does in the tropical regions of South America and Africa; large areas of original forest have been cut down to make way for cultivation. It is only on the islands of the Sunda shelf that substantial vestiges of the original flora and fauna remain. In the rainforest, the tops of the trees form a compact canopy which is broken only here and there by the emerging tip of a palm-tree. Beneath this roof of vegetation there is a dense tangle of creepers, epiphytes and bamboos.

The Monsoons

Monsoons are the fundamental climatic feature of the regions surrounding the Indian Ocean. They are periodical winds which change direction seasonally according to changes in the interaction between warm air from the tropics and the cooling influence of the vast oceanic areas to the south and the continental land-mass to the north. In summer, as a result of the intense heating effect, the mountain-chains of Asia become areas of low pressure (cyclonic zones) which attract warm, humid winds from the sea *(see map, left)*. This time of the year is accordingly humid with frequent rain and is particularly favourable both for natural vegetation and for agriculture. In winter, the continental land-mass cools more rapidly than the ocean water and it therefore becomes an area of high pressure (anticyclonic zone) so that cold, dry winds sweep out to the sea *(see map, right)*.

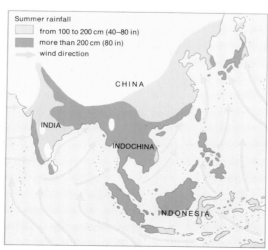

Summer rainfall
□ from 100 to 200 cm (40–80 in)
■ more than 200 cm (80 in)
→ wind direction

CHINA

INDIA

INDOCHINA

INDONESIA

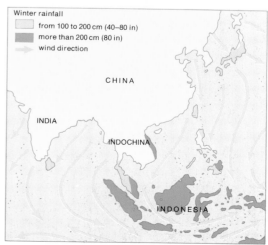

Winter rainfall
□ from 100 to 200 cm (40–80 in)
■ more than 200 cm (80 in)
→ wind direction

CHINA

INDIA

INDOCHINA

INDONESIA

The Indian Triangle
The triangular peninsula formed by India and Pakistan is a transition zone with respect to its animal inhabitants. Some tens of millions of years ago, India was a vast island slowly moving northwards over the Earth's surface. Whilst India was subsequently in the process of fusing with Asia, Africa was also closing the gap with Europe and the Near East, approximately fifteen million years ago. This set the stage for a massive exchange of animals between land-masses. Rhinoceros, giraffe and canids (members of the dog family) spread westward, while elephants and lions moved towards India. Although the details of this animal exodus remain unclear, it is certain that over the last three thousand years migrations of animals to the west have been barred by the great Indian desert, which is a formidable obstacle for any animal species not specifically adapted for arid conditions. Most of the present animal inhabitants of the rainforest are related to Malayan and Indochinese species. The top photograph *(above)* shows a general view of the rainforest, while the other shows a banyan or Indian fig-tree (a typical tree species of this area). *Below:* A typical sky just prior to the monsoon period.

India

Three zones are distinguished in India: the high, central Deccan plateau, characterized by a temperate climate and (originally) by extensive forest cover; the arid plain of the Indus; and the rainforest zone of Assam. The major part of the region situated along the Indus is desert-like and is inhabited primarily by animals of Palaearctic type. Typical examples are the nilgau (bluebuck), the Indian antelope (blackbuck), the Persian gazelle, porcupines, wild boar, shrews and a variety of bat species. Among the predators, the most impressive carnivore is the leopard, but there is also the Bengal fox and several mongoose species. In winter, this zone is visited by numerous migratory bird species which fly down from Siberia: ducks, geese, pelicans and cranes. The non-migratory bird species of the Indus valley include moorhens, purple gallinules, spoonbills and egrets. Assam, in direct contrast with the arid plain of the Indus, is one of the richest regions of India in terms of plant species, since the humid climate favours the growth of dense forest. Orchids are particularly diverse, while azaleas and rhododendrons are also common. Unfortunately, forest clearance and uncontrolled burning have considerably modified the original virgin forest of Assam, and this modification has in turn had repercussions for the fauna, with many species now on the verge of extinction. Nevertheless, it is still possible to find a number of cervid species (members of the deer family), such as axis deer, muntjac, sika and sambar. The Indian jungle is one of the most prolific areas of the world in terms of bird species, notably for gallinaceous birds: peafowl, partridges, pheasants and quail.

The Vegetation of India
Because of its extensive spread in latitude, India possesses a very diverse flora. Rattan palms *(Calamus)*, bamboos, mangrove-trees, fig-trees, ferns and a variety of grasses combine to produce a wealth of different habitats, ranging from evergreen forest through jungle to prairie, according to prevailing rainfall levels.

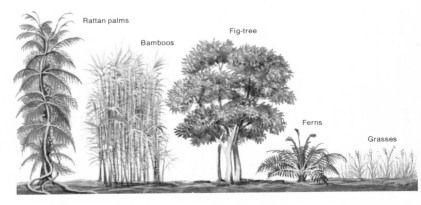

Rattan palms

Bamboos

Fig-tree

Ferns

Grasses

Animals of the Jungle

The symbol of India's animal-life is incontestably the tiger, which seeks out cool, shaded places close to water. But this big cat is becoming increasingly rare, whereas leopards are still abundant and will not hesitate to approach villages to raid their domestic animals. The sloth bear (or Indian bear) is also common in the jungle, where it feeds upon fruits, honey and insects. This bear's exceptionally well-developed sense of smell permits it to detect the presence of a larva at the end of a narrow gallery 60 cm (24 in) in length. Jackals, hyenas and vultures share out the remains of kills made by the big carnivores, sometimes even wresting the prey from their grasp. In order to avoid such poaching, tigers often hide their prey under a pile of leaves, while leopards—which are more agile—carry their prey into trees. Small Bengal foxes, which are still quite common in central India, feed on reptiles, frogs and insects. On the other hand, the wolves which figured so prominently in Kipling's tales have become extremely rare. Macaques and langurs, the two kinds of monkeys to be found in India, are widespread both in the countryside and in villages, where everyone tolerates their presence despite the damage they cause.

The Indian Elephant
The Indian elephant, smaller and more docile than its African relative, occurs throughout the forests of Asia.

Water Buffalo
These even-tempered bovids spend a large part of the day wallowing in water and feeding on marsh-living plant species.

A Victim of Superstition
Like all the other Asiatic rhinoceroses, the Indian species is threatened by extinction. The rhinoceros has long been hunted for its horn, which according to popular superstition is supposed to have valuable medicinal and aphrodisiac properties.

Poisonous Snakes
India is famous for its snakes, the most famous of all being the cobra (right). The powerful venom of the cobra is largely composed of neurotoxins, which act directly on the central nervous system; haemolysin, which destroys red blood cells; and anticoagulants, which facilitate the spread of the poison through the victim's body. Other Indian snakes belong to the viper family (for instance the big Russell's viper) or to the pit-viper family. The venom of these snakes destroys the walls of blood-vessels and provokes widespread clotting.

The Plain of the Ganges
The photograph (above) shows the appearance of the Ganges plain when it is flooded by the waters of this great river. Flooding of the Ganges occurs with great frequency and the floods are often disastrous in the south-east sector of the Indo-Gangetic plain. They result from the heavy rains which fall during the monsoon period, amounting to as much as 10 metres (33 feet) of rain per year in regions such as Assam. Added to this is the fact that the Ganges originates in the Himalayas, with several peaks exceeding 8,000 metres (26,000 feet) in height. The extensive ice layer adds considerably to the flow of water feeding the Ganges. Finally, it must be remembered that the Ganges, after it has crossed the heavily populated Indian plain, is joined by the Brahmaputra, which also flows down from the Himalayas. The diagram (right) shows the massive mountain-ridge formed by the Himalayas, separating the fertile Indo-Gangetic plain with its heavy monsoon rainfall from the arid high plateaux of Tibet.

INDIAN SUBCONTINENT

Mount EVEREST

Calcutta

Ganges

Ganges plain

HIMALAYAS

Brahmaputra

Lhasa

TIBET

Dacca

Brahmaputra

8 000 — metres
6 000
4 000
2 000
0

Mount EVEREST

Brahmaputra

Lhasa

Ganges

Mangrove Forest

Mangrove forest, which characteristically develops as a belt along tidal shorelines, is a typically tropical habitat found in Africa, the Americas, Australia and Asia. In the muddy water among the tangled roots there is a rich animal life including numerous bivalve mollusc species, a variety of crabs and shrimps of differing sizes and many different species of fish. This profusion of animals attracts a whole spectrum of predators. Monitor lizards, long reptiles which can reach up to two metres (six feet) in length, pursue their prey both in the water (where they are able to swim perfectly) and in the trees, (where they can move with equal agility). Marine crocodiles, which will move out to the open sea and can be a serious menace for fishermen, make their nests in estuaries and take shelter in the mangrove swamps. Several sea-snakes, all extremely poisonous to man, add to the dangers of this coastal forest zone. The predators also include a number of bird species, such as the sea-eagle and the Brahminy kite. Adjutant storks of the genus *Leptoptilos* slowly wade through the swamps, stabbing at any fish within range with their long beaks, while brightly-coloured kingfishers dive into the water to seize their prey. Crows, by contrast, largely feed on dead fish and other carrion washed up by the waves. Swallows and various flycatcher species perform a useful service by hunting the mosquitoes and other biting flies which infest the region. Wild boar are also occasional visitors to the mangrove swamps, attracted by the abundance of crustaceans and molluscs which are favoured food items; and there are fish-eating otters as well. A monkey species—the crab-eating macaque—has also become specialized for feeding on crabs. After a macaque has spotted a crab taking refuge in the sand, it will remain immobile until the crab reappears to be seized and eaten. The proboscis monkey, with its remarkably well-developed nose, is another typical monkey inhabitant of mangrove forest; but this species feeds only on fruits and leaves. Finally, there is the unusual dugong, which looks rather like a seal in some respects but is quite different in being purely herbivorous, feeding exclusively on mangrove leaves. Mermaid legends may well have arisen because of the persistent melancholy calling of the dugong and the female's habit of holding her offspring against her human-like breasts.

A Natural Refuge
The dense tangle formed by the mangrove roots and the surrounding mud, which becomes progressively drier as one moves away from the shore, provides a very suitable hiding-place for many little burrowing animals, including several crab species and certain ants.

Geographical Distribution of Mangrove Forests
Long stretches of mangrove swamp can be found along all the coasts bordering tropical seas. Two major zones are recognized, one in the Indo-Pacific region *(map, left)* and the other in the Atlantic region (West Africa and the eastern coastline of the New World).

The Flying Fox
Bats, the only mammals able to fly, are distributed throughout the temperate and tropical areas of the world. In tropical regions, the bat species include a number of large-bodied megachiropterans, commonly known as "fruit-bats" or "flying foxes" *(left)*. The distinctive shape of the head in such bats is, indeed, reminiscent of a fox. These mammals are only active at night and they spend the day hanging (head downwards) from the branches of trees or in caves. Fruit-bats are particularly common in mangrove swamps, which provide them with a safe refuge. They feed on fruits, which they are able to find even in very dim light conditions. Unlike the other bats (microchiropterans), however, they rely on sight rather than on echo-location. Microchiropteran bats emit high frequency sound pulses which are almost entirely inaudible to man and which are reflected by any objects which they strike. The echoes are picked up by the bat's ear, which is specially adapted to perceive ultrasonic waves. Whenever they are emitting sound-pulses in this way, bats close their ears and then re-open them to pick up the echoes. This rapid opening and closing of the ear is made possible by the development of a special small muscle.

The Composition of Mangrove Forest

The coastal forests of hot countries are characterized by the predominance of plant species with well-developed adventitious roots (e.g. prop roots) and pneumatophores (upward-growing respiratory roots) as a special adaptation for muddy soils lacking in air. There are a number of recognizable plant associations which occur as a series of zones relating to the tidal flow. The commonest trees are mangroves belonging to the genera *Rhizophora* (black mangrove), *Bruguiera* (red mangrove) and *Sonneratia* (white mangrove), while the characteristic small plants are rushes and glassworts. In the area of the mangrove swamp which is always submerged, algae and pondweed occur. An environment of such diversity obviously provides a large number of ecological niches and thus permits a wide variety of animal species to live there.

Silver leaf-monkey

Proboscis monkey

Painted stork

Red mangroves

Black mangroves

White mangroves

Krait

Crab-eating frog

Mudskipper

Archer-fish

Animals of the Mangrove Swamp

Characteristic inhabitants of the mangrove swamps of South-East Asia are the proboscis monkeys, which are at home not only in the trees but also swimming among the protruding mangrove roots. Fishing cats, which feed on molluscs as well as fish, are also quite common, as are long-billed ibis, strange fish which are able to survive out of the water, and a variety of crustacean species. Indian pythons and mangrove snakes are found lurking among the mangrove roots. Both these species feed on birds, but only the latter is poisonous.

Mangrove Roots

White mangroves (*Sonneratia*) have roots with pointed tips which grow upwards and stick out above the surface. These roots are thought to serve a respiratory function but are regularly submerged at high tide.

The Fiddler Crab

Among the black mangrove trees *(Rhizophora)*, not far from the shore, a typical inhabitant is a small crab with bright red coloration and an enormous pincer which is continually moved back and forth. This "fiddler" crab *(below)* belongs to the genus *Uca* and its large pincer (which is a feature of males only) is not used for fighting or feeding but for courtship. The male signals with his pincer to attract the female. At the slightest sign of danger, these small crustaceans are able to dig into the mud and disappear within a few seconds.

The Fish that Walks

One of the most unusual sights in the mangrove swamp is that of dozens of little fish which jump out of the water, crawl along the ground and climb into trees, where they attach themselves with a special sucker formed from the fused pelvic fins. These are mudskippers and they move along by "walking" on their pectoral fins.

The Archer-fish

Another bizarre inhabitant of the mangrove swamp is a small fish known as the "archer-fish" because of its curious method of preying on insects *(above)*. The archer-fish is equipped with a groove running to the tip of its mouth. Water droplets are projected rapidly along this canal just like arrows to dislodge insects resting on plants.

Sonneratia

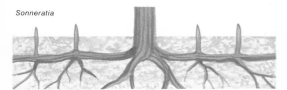

Rhizophora

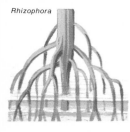

Black mangroves have secondary roots which grow out from the trunk and branches. These roots help to anchor the tree in the soil and in fact play an important part in the development of the mangrove swamp substrate, as they hold back detritus.

Bruguiera

Red mangroves have creeping roots which penetrate the soil only superficially. Although the upper zone occupied by red mangroves has a deeper soil, there is still a high salt content and complete saturation with water.

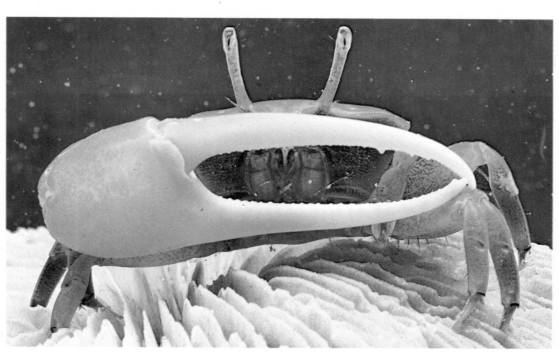

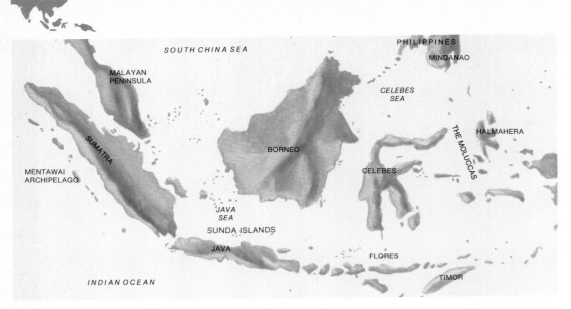

SOUTH CHINA SEA

MALAYAN
PENINSULA

PHILIPPINES
MINDANAO

SUMATRA

MENTAWAI
ARCHIPELAGO

BORNEO

CELEBES
SEA

CELEBES

HALMAHERA

THE MOLUCCAS

JAVA
SEA

SUNDA ISLANDS

JAVA

FLORES

TIMOR

INDIAN OCEAN

The Islands of the Sunda Shelf
The Indonesian archipelago constitutes one part of the great island arc of South-East Asia. These are relatively recent islands in geological terms; they were formed during the Tertiary (Cenozoic) era at the same time as the Alps and the Himalayas. The fact that this region is still geologically young and has not yet reached a stable, quiescent state is witnessed by the presence of intense seismic activity and a large number of volcanoes. Java alone contains more than 120 volcanoes, with fifteen still active.

Ten Thousand Islands

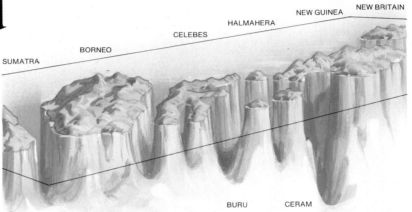

SUMATRA
BORNEO
CELEBES
HALMAHERA
NEW GUINEA
NEW BRITAIN
BURU
CERAM

Unexpected Precipices
A vertical section through the Indian Ocean (left) reveals an unusual aspect of the Indonesian archipelago; there is a deep trench which reaches a depth of 7,450 metres (24,590 feet) south of Java. The Java trench represents the line of contact between the Australian tectonic plate in the south-west and the Eurasian plate in the north-east. The former is slowly sliding beneath the latter through the process of subduction which produced the trench in the first place.

To the south of the Indochina peninsula, Asia is fragmented into a series of islands numbering about ten thousand altogether, stretching as far as Australia and New Guinea. However, despite the resulting close proximity between the two geographical regions (Asia and Australasia), their faunas have still remained very distinct. In fact, Asian animal species have spread through the large islands of the Sunda shelf, whereas Australian species migrated to New Guinea and Tasmania. This phenomenon is explained by the particular structure of the sea floor in this area, which presents an insurmountable barrier. The Malay archipelago has a hot, humid climate with temperatures averaging 30°C and very little difference between the coldest and hottest months. In addition, there is high rainfall which is spread fairly well throughout the year. Unfortunately, the geological situation is not so stable, for there are about five hundred volcanoes in the area, of which close to a hundred are still active and continually menace these richly forested islands. In addition to vast expanses of mangrove swamps and coastal forests containing screw-pines, hibiscus and palm-trees, the islands of South-East Asia are largely covered with rainforest. The undergrowth of the rainforest is relatively sparse, but access is limited because violent rain-storms generate great swamps where leeches abound. The flooded soil does not provide a solid foundation for the large trees and these are often brought crashing down during storms, carrying with them everything in their immediate vicinity. Such fallen trees are at once attacked by hordes of termites, by the larvae of wood-boring insects, by mushrooms, by moulds and by bacteria which live on decayed matter. As a general rule, the animal species found in these forests are the same as those which inhabit the mainland rainforest of South-East Asia.

The Vegetation
In the driest areas, there are small patches of screw-pines, which are characterized by the presence of spiny leaves and fruits resembling pineapples. Sago-palms and hibiscus (left) are found in more humid areas. Sago-palms (peculiar trees closely related to palms) are survivors from the Mesozoic era and are virtually confined to this region. But the best-known plant association of the Malay archipelago is the rainforest: an extremely complex assemblage of plant species, including a very large number of different species.

Terrace Cultivation
Because of the differing climatic influences which come to bear, involving both the proximity of the great continental land mass to the north and the presence of Australia in the south, the rainfall regime of the Malayan archipelago varies greatly from one island to another, though rainfall is actually quite heavy everywhere. Rainfall varies from 200 to 400 cm (80 to 160 in) per year and produces luxuriant plant growth, particularly on the islands which are most ancient in geological terms (Sumatra and Borneo), where the soil is deep and rich in nutrients. In addition to the characteristic mangrove forests, there are zones where agricultural development is considerable. Because of the hilly terrain, the most widespread system of cultivation for maize, tobacco, bananas and rice is the terrace system. Right: Terraces in Bali.

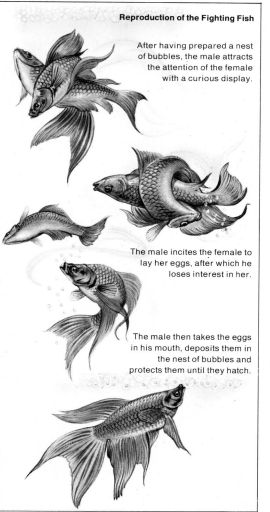

Reproduction of the Fighting Fish

After having prepared a nest of bubbles, the male attracts the attention of the female with a curious display.

The male incites the female to lay her eggs, after which he loses interest in her.

The male then takes the eggs in his mouth, deposits them in the nest of bubbles and protects them until they hatch.

Gliding Animals
The Malay archipelago includes a large number of gliding animals, including the flying squirrel *(above)* and the flying snake *(below)*. The latter has long, mobile ribs which are extended to create a hollow along the belly, thus increasing the ventral surface area and allowing the snake to glide down from the top of a tree without harm.

A Disappearing Ape
The orang-utan *(above)*, now confined to certain forest areas of Borneo and Sumatra, is very discreet and wary of man, but it is nevertheless threatened with extinction because of hunting and, especially, habitat destruction.
Below: A snake-bird, or anhinga.

45

Life in the Rainforest

In South Asia the rainforest no longer forms a continuous zone because large areas have been transformed by human activity. But the mountain slopes of Indochina still have dense forest cover. Typical tropical rainforest is found up to altitudes of 1,500 metres (5,000 feet) on the mainland, and it is widespread on the islands of the Sunda shelf. Any food which is available in this forest is largely to be found in the trees themselves, since the undergrowth layer is dark and impoverished. Because of the year-round uniformity of the climate, flowers and fruits are available virtually throughout the year, but individual species have different times of flowering and fruiting. Most of the animals have therefore become adapted for arboreal life, with a resultant slow evolutionary change in their body form. All the larger arboreal animals share a slender body form, a colour pattern which blends with the foliage and the ability to leap from one branch to another. The rainforest is also a favourable environment for small-bodied animals, which can more easily exploit the arboreal vegetation. The Indo-Malayan region provides a home for an unbelievably large number of insects, some of which are enormous. Examples are the tropical wasp *(Vespa tropica)*, with a length of some 10 cm (4 in); a giant bee *(Apis dorsata)*, which constructs huge honeycombs high up in the trees;

the carpenter ant, which exceeds 2 cm (1 in) in length; and the weaver-ant, which uses the sticky secretion of its larva to glue together the edges of the leaves that form its nest. The butterflies can also be very big and they are often very brightly coloured. Cockchafers and beetles (including the famous rhinoceros beetle) fly around in search of flowers so that they can imbibe the nectar and, incidentally, play their part in the pollination process. In addition, there are stick-insects, leaf-insects, dragonflies, praying mantises with a variety of unusual shapes, and a multitude of spiders, millipedes and scorpions. Although birds are quite common, they generally remain virtually hidden in the burgeoning vegetation and it is therefore difficult to observe them. The undergrowth, which is impoverished and provides little in the way of shelter, yields only limited amounts of food and is generally unsuitable for large mammals. In contrast to the situation out in the savannah, hoofed mammals (ungulates) do not form large herds here; the largest of them, such as the elephant and the buffalo, tend to be rather solitary in habits. The small mammals lead a discreet, nocturnal existence and are preyed upon by civets and mongooses. Larger herbivorous mammals are, in turn, attacked by ferocious big cats such as the Bengal tiger and the leopard.

Black gibbon
This lesser ape *(Hylobates concolor)* is found in the forests of Indochina.

Forest Acrobats
Gibbons, or lesser apes, are born acrobats. They can move around rapidly in the trees by brachiating—swinging from one arm to another beneath branches. When doing this, they can make leaps of as much as 10 metres (33 feet). They live in groups composed of three to five individuals and prefer dense forest, either on mountain slopes or in lowland areas. Their diet is composed mainly of fruits, with some leaves and the occasional insect, egg or young bird.

The Venus' Fly-trap
Like the pitcher-plant, the Venus' fly-trap is an insectivorous plant whose leaves are converted into traps. The fly-trap is a perennial plant which grows on peaty soils. The insect-capture mechanism differs markedly from that of the pitcher-plant, though the end result is the same, namely the provision of nitrogenous material to the plant. The leaves of the fly-trap consist of two flat lobes which are fringed with spikes. As soon as an insect touches the sensitive hairs in the centre of a lobe, the leaf folds together and traps the prey.

The Reproductive Cycle of Ferns
The reproductive cycle of the fern *(above)* is very complex. The lower surface of the sporophyte is covered with numerous sporangia in which the spores ripen. In Spring, the sporangium opens and any spore which falls to the ground produces a prothallus which anchors itself to the ground with special filaments and is able to live independently because of its ability to photosynthesise. The prothallus differentiates into antheridia and archegonia, which produce, respectively, the male gametes (antherozoids) and the female gametes (oospheres). The oosphere is fertilized by an antherozoid and then develops to produce a new fern which at first lives at the expense of the prothallus and then later develops the structures necessary for independent existence.

Sorus

Sporangium

Spores

Rhizome

Prothallus

Young plant

Spore

Pitcher-plants
The pitcher-plant *(Nepenthes)* is a climbing plant which possesses brightly-coloured tubular leaves (ascidia). They secrete a nectar which attracts insects. The internal surface of the ascidium is smooth and any insect which settles on it falls to the bottom of the ''pitcher'' where it is digested by enzymes. The section through a ''pitcher'' *(right)* shows the viscous liquid into which the insect falls.

Nectary-bearing lid

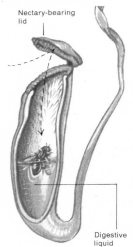

Digestive liquid

The white-handed gibbon
This gibbon *(Hylobates lar)* is also known as the lar gibbon.

The siamang
This is the largest of the lesser apes *(Symphalangus syndactylus)* and lives in the densest forest areas up to an altitude of 1,500 metres (5,000 feet).

The hoolock gibbon
Whereas the male of this species *(Hylobates hoolock)* is always black, the female is somewhat variable in colour, though typically brown.

Orchids
The orchid family, which contains about twenty thousand species belonging to seven hundred genera, is the largest plant-group and is primarily distributed in hot and temperate zones of the world. These are monocotyledons in which there is a specially modified single sepal (labellum), located ventrally, to attract insects for pollination.

Epiphytes *(below)*
Epiphytes—plants which grow on other plants—are actually ground-adapted species in some cases, with an underground root, often ending in a tuber. Other species have aerial roots capable of photosynthesis; yet others have tendrils which are entwined around trees.

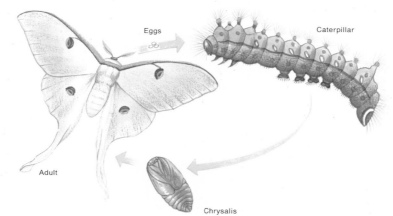

Eggs

Caterpillar

Adult

Chrysalis

Arboreal Predators
Predators, which are the most prominent tree-living species to be found in tropical forest, also show special adaptations for arboreal life, as in the case of the leopard *(above)*.

Metamorphosis
Metamorphosis refers to the radical restructuring of the body plan of any given animal, such as the moth *Actias selene (above)*. At each passage from one phase to the next, a metamorphosing insect metabolises its own body, breaking down its old cells to produce simpler compounds for the construction of new tissue. This phenomenon is governed by glands which secrete hormones under the influence of the central nervous system. In the butterfly, there is a transition from a mobile, actively feeding stage (caterpillar) to an inactive, static phase (nymph), prior to attainment of the adult phase in which reproduction becomes possible. The function of the larva is therefore that of obtaining food substances which were not available in the egg to permit production of the eventual adult stage. Development of the sex organs depends on the balance between the juvenile hormone (neotenine), which is more abundant in the larva, and the moulting hormone (ecdysone), which is present in greater amounts in the adult.

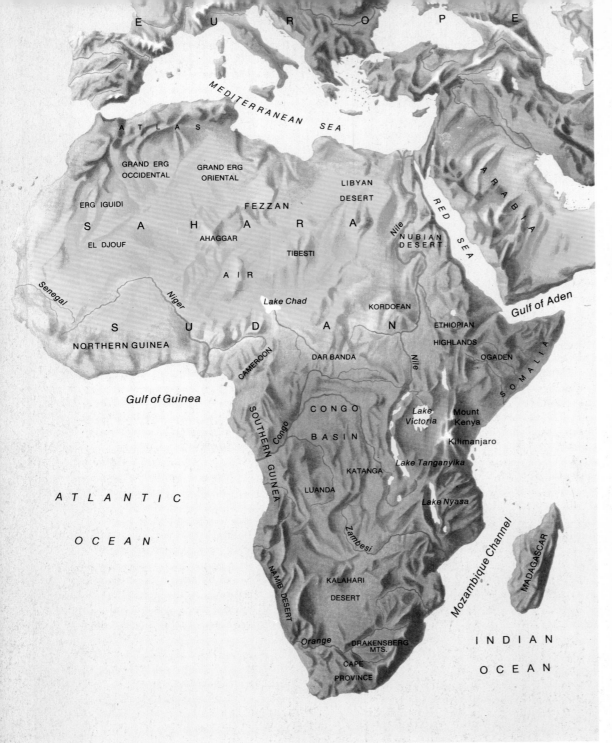

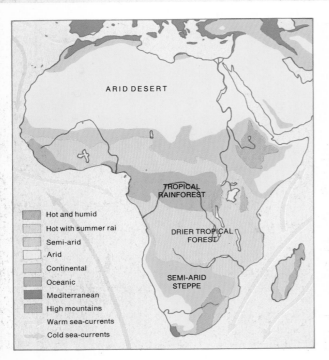

AFRICA

Africa, the second largest continent on Earth, has a total area exceeding 30 million square kilometres (12 million square miles). Because of its particular geographical location, this continent contains an entire spectrum of climatic conditions. In fact, there is a mirror-image effect, with a subtropical climate in the north giving way first to desert, then to savannah and finally to rainforest, after which the sequence is reversed. These vegetational types form clearly demarcated zones, each characterized by its own large assemblage of plant and animal species. South of the narrow, subtropical Mediterranean belt stretches the great Sahara desert, covering a total distance of almost 5,000 kilometres (3,000 miles) from the Atlantic Ocean to the Red Sea. Beyond the desert lies a grassland zone which gradually gives way to wooded savannah as rainfall increases. Finally, there is the extensive equatorial rainforest belt which in extent is second only to the vast Amazonian rainforest. The African equatorial forest contains an enormous variety of animal species, ranging from a multitude of insects of all shapes and sizes to large mammals such as buffalo and elephants.

National Parks and Reserves
The savannah areas of Africa are regarded as the last refuges for large numbers of animal species. Vast national parks have been created in the savannah regions in order to provide official protection for certain animals such as rhinoceros, eland and giraffe (above).

Unlike Asia or the New World, Africa does not have any really large-scale mountain-chains. Instead, there is a series of great plateaux on which there are isolated peaks here and there, the highest being Kilimanjaro and Mount Kenya in the east. East Africa also has a number of special features such as the immense Ngoro-Ngoro crater, which was probably formed by the impact of a huge meteorite, and the majestic Rift Valley, an enormous fissure in the Earth's crust running parallel to the Red Sea and including a number of lakes, some of gigantic dimensions.

The Tree Senecios of Ruwenzori
Tree senecios—arborescent shrubs related to groundsels which are characteristic of high mountain areas in Africa—can survive at altitudes up to more than 4,000 metres (13,000 feet) because of the general tropical climate of the region. The photograph (right) shows in the background Mount Baker, one of the peaks of the Ruwenzori group.

The African Deserts

The Sahara, a desert extending more than 4,800 kilometres (3,000 miles) from west to east and 1,600 kilometres (1,000 miles) from north to south, constitutes a massive, arid area below the narrow strip of Mediterranean coastal vegetation. But the landscape of the Sahara is actually quite varied; mountain peaks and massifs are interspersed with vast stony plains and great expanses of sand-dunes. The temperature difference between day and night is very large and gives rise to violent winds which carry dust and sand before them, producing a continual upheaval of the desert surface. Life in the desert is accordingly very difficult and the constant need for water is a dominant feature. Thickets of tamarisk, oleander and acacias grow along the wadis—rocky watercourses which are fed by water running down from the Atlas Mountains. The same plant species are also found along the southern boundary of the Sahara. However, in the core of the desert perennial plants are very rare. Animals, of course, depend upon the few plants which

are present, either as a direct source of food (e.g. for certain insects and herbivorous mammals) or indirectly as the food of their prey (as with carnivorous species such as toads and reptiles). Desert plants are therefore exposed to a double menace: from persistent aridity and from hungry animals. In order to protect themselves from the latter, they are often covered with spines, shielded by a waxy coat or simply defended by noxious substances. A waxy coat and spines also assist in water-retention.

There is another expanse of desert in the south of the African continent. The coastal Namib desert is a desolate area characterized by huge expanses of stones alternating with towering sand-dunes some 150 to 300 metres (500 to 1,000 feet) in height. In this desert, vegetation is extremely sparse and consists only of a few acacia and tamarisk species along with thickets of certain very hardy plants. The Kalahari desert, by contrast, is rather richer in vegetation and its sand-dunes range in colour from bright pink to red.

The High Atlas (Morocco)

The complex Atlas mountain-chain, which is a geological contemporary of the Alps, can be considered as the north-western boundary of the vast Sahara plateau. The Atlas mountains in fact contribute to the arid conditions of the desert, since they act as a barrier to winds from the Mediterranean Sea, but they are at the same time vital to the maintenance of oases. This is because rainwater penetrates the fractured rocks of the Moroccan chain and then flows between two impermeable rock layers, lying about a kilometre (more than half a mile) underground before re-emerging to form oases where special geological structures of the Sahara (geosynclines) dictate.

Oases

Oases are primarily located in depressions and they often lie below sea-level. The typical vegetation surrounding them includes date-palms, pistachio-trees, spiny bushes bearing berries (such as *Lycium*), and grasses of the genus *Panicum*.

Sand Deserts

Sand-dunes, which can reach heights of up to 300 metres (1,000 feet) in the desert are generated by wind action. Sand desert, also known as *erg*, consists of a chaotic assemblage of dunes which continually change shape.

Stone Deserts

In places, wind action removes the finer fragments produced by the disintegration of rocks, leaving behind a carpet of stones with outcrops of solid rock here and there. The uniformity of such stony desert areas is occasionally interrupted by heavily eroded peaks, as in the Hoggar *(above left)*, which are volcanic in origin and may reach 3,000 metres (10,000 feet) in altitude. Such peaks are vestiges of the basalt outpourings which took place during the Tertiary and Quaternary eras, basalt being more resistant to fragmentation than the surrounding rocks.

Animals of the Sahara

In erg regions, there is a much greater density and variety of animals than in rocky or stony desert areas, since the mobile sandy terrain provides greater opportunities for survival. Fennecs (desert foxes) and gerbils dig burrows and nest-cavities in the sand and take refuge there during the hot part of the day. There is a surprising variety of animal species in the Sahara. It is even possible to find toads, which belong to the amphibian group and depend upon water for the development of their eggs. As soon as temporary pools are formed during violent storms, they are occupied by toads which have a specially rapid reproductive cycle (eggs, tadpoles, adults) which is completed before the water dries up. In order to survive until the next fall of rain, these amphibians return to a burrow of some kind and become torpid. In this state, they breathe very slowly and are able to lose up to sixty per cent of their water without succumbing. Large mammals, such as the oryx *(above left)* and the dromedary *(left)* can also be found in the erg.

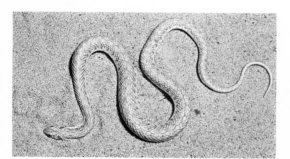

Among the Hot Dunes of the Namib

Despite the extremely arid conditions, the sand-dunes of the Namib desert have a rich animal life. Various darkling beetle species belonging to the family Tenebrionidae *(above right)* are particularly common. These beetles have very hard carapaces which prevent them from becoming dessicated. They feed upon particles of organic matter carried by the wind, often from some distance away. The Namib desert is also inhabited by spiders, scorpions and several snake species. The tiny web-footed gecko *(above left)* owes its name to the special webbing on its hands and feet which permits it to move rapidly over the sand-dunes. In areas where there is a gravel substrate, it is also possible to find hares, antelopes, hyraxes, genets, hyaenas and a variety of birds (e.g. weaver-birds, falcons and vultures).

Snakes and Lizards

The animals which are best adapted for desert life are incontestably the reptiles, which satisfy their water requirements by extracting the body fluids of their prey. Moreover, thanks to the impermeability of their skin and the almost solid state of the urine they produce, their water-loss is reduced to a minimum. Desert-living snakes, such as the sand viper (above), can bury themselves in the sand within a few seconds, closing off their nostrils with a system of valves so that they do not suffocate. The major enemies of the reptiles are various birds of prey. Bell's Dabb lizard *(above left)* will thrash out with its tail to defend itself against predators such as eagles, falcons and goshawks. *Left:* Dark chanting goshawk.

The Southern Deserts

This small map shows the position of the two main deserts of southern Africa, the Namib along the west coast and the more centrally located Kalahari.

The Namib

The coastal band of desert constituting the Namib is made up of gravel in the north and sand in the south. The only reliable source of humidity is the night-time fog which is produced by the cold Benguela current and which supports the sparse vegetation on which the animals depend.

The Kalahari

In contrast to the extremely arid Namib desert, the Kalahari is quite well supplied with vegetation, to the extent that it looks more like rather dry grassland than a desert in the strict sense. Several large mammals, including kudu, gemsbok and wildebeest, lead a nomadic existence there and obtain the water they need for survival from various plant species.

The Vegetation

Aloes, which are perennial plants belonging to the family Liliaceae, are characteristic inhabitants of the African deserts, well adapted to dry conditions and often reaching several metres in height. *Below:* Two aloe species found in the Kalahari.

Namib   Kalahari

The Namib and Kalahari Deserts

The diagram *(above)* illustrates the location of the two great deserts of southern Africa. The vast Kalahari, which is almost entirely included within the state of Botswana, lies at an altitude of about 1,000 metres (3,300 feet). Its geological structure consists of a very ancient crystalline foundation (of Precambrian origin) covered by more recent marine sediments, and it is almost entirely uniform throughout. The mountain-ridges which surround it, including the high plateau which separates the Kalahari from the Namib, block off any influx of humid winds so that the region is very dry, with less than 25 cm (10 in) of rain a year. Rain is also very rare in the Namib desert, despite its location parallel to the Atlantic Ocean, for the winds pass along the coastline and never carry water vapour inland.

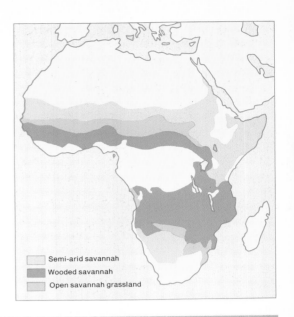

Steppe, Savannah and Prairie

The first vegetation zone to be encountered just south of the Sahara is dry steppe. The plants in this zone produce large quantities of seeds and these provide food for a large number of seed-eating birds and small rodents. The majority of the large mammals in this area (mainly oryx and gazelles) lead a nomadic existence, continually seeking out fresh grazing- and watering-places. Because these mammals are so agile, only the cheetah is able to catch them. Further south, rainfall progressively increases and the steppe zone gradually gives way to savannah, which has a richer plant growth and even the occasional tree. In Guinea, trees are taller and more frequent, and this wooded savannah eventually leads on to tropical forest. Savannah, unlike the forest, is occupied by animals specialized for feeding on grass and seeds, and any nests are constructed on the ground. It is generally accepted that human activities and periodic fires have played a decisive role in the formation of the savannah, but there is also considerable modification by the animal community. Elephants, in particular, destroy vast areas of tree cover and in this way prepare the terrain for the development of prairies. South of the tropical rainforest of the Congo basin, there is once again a vast savannah belt which stretches from one side of Africa to the other. This is, in fact, one of the regions of Africa which has been least disrupted by man, since the presence of the tsetse fly (the carrier of sleeping sickness) has effectively prevented the establishment of herds of cattle there. This region, which is characterized by tall grass and sparsely distributed trees, is fundamentally different from the savannah zone found north of the equator because of the timing of the rainy season. When conditions are driest in the north, there is heavy rainfall in the south and the resulting flush of vegetation attracts large numbers of migratory birds which fly down either from the northern savannah belt or from Eurasia. Otherwise, however, the animal species found in the southern savannah zone are closely related to those found in the savannah and prairie zones north of the equator.

The Landscape

Many savannah areas are the products of forest degradation, accompanied by marked impoverishment of the soils, which range from the ferrugineous (iron-impregnated) earths of the tropical zone to laterites (a type of surface clay) in the proper sense, according to the degree of impoverishment.

The Vegetation

The dry season is a resting period for the seasonally adapted plant formations, which change in appearance according to the time of the year. During the driest part of the year, the trees lose their leaves and the landscape has an air of desolation.

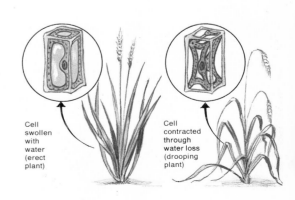

Osmosis

Osmosis is a phenomenon which occurs when two aqueous solutions are separated by a semi-permeable membrane, such as a cell membrane, which will permit the passage of the solvent (water) but not of the substances dissolved in it. Water flows from the weaker to the stronger solution until equilibrium is reached due to the hydrostatic force generated by the cell wall.

Cell swollen with water (erect plant)

Cell contracted through water loss (drooping plant)

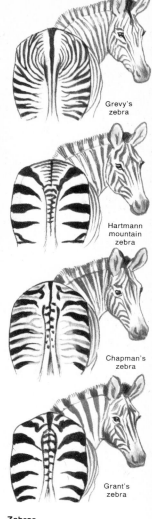

Grevy's zebra

Hartmann mountain zebra

Chapman's zebra

Grant's zebra

Herbivores and Seed-eaters

Most of the mammals in the prairie are grazers. Antelopes, such as impala *(left)*, zebras *(above)* and gazelles feed on grass at different growth stages and thus avoid competition. Zebras feed on tall grass and when a particular area is no longer suitable they move on to new grazing-grounds, to be replaced first by antelopes and then by gazelles. When there is nothing left on the ground but dry stalks, the blesbok (larger-bodied antelopes) take their turn and virtually denude the prairie, thus limiting the danger of fires and setting the scene for the growth of new grass. In bushland areas, close to the river courses, there are kob, impala, warthog, eland and buffalo. Giraffes *(above left)* feed on acacia leaves and are therefore found in wooded savannah or in small thickets close to water. Since seeds are abundantly available, this habitat also provides a home for numerous seed-eating birds, the commonest being weaver-birds, weaver finches, guinea-fowl and pintailed sandgrouse.

Rhinoceroses

Although the two African rhinoceros species are both herbivores, they do not compete with one another for food. The black rhinoceros *(left)* feeds on the tough leaves of bushes, while the white rhinoceros *(right)* grazes out on the grassland.

Zebras

All the zebra species have striped coats and this makes it difficult to distinguish them. However, the pattern of stripes on the hindquarters *(above)* does permit clear identification.

The Predators

The large numbers of herbivores inhabiting the prairie and savannah regions naturally attract many predators, the most important being the lions, cheetahs, hunting-dogs and, to a lesser extent, leopards. Lions live in social groups and will hunt any of the savannah herbivores, though they concentrate on zebras and large-bodied antelopes. The hunting techniques used by the predators vary considerably. The ferocious African hunting-dogs hunt in compact packs, pursuing their prey to the point of exhaustion, whereas the cheetah makes surprise attacks at great speed on gazelles or small antelopes which are killed by a bite through the throat. The number of predators in the savannah is further increased by a variety of snake species, including the deadly mambas, and a number of birds of prey. Among the latter, there is one species of particular interest, the secretary bird, which is terrestrial and feeds on several different snake species. Both snakes and rodents are preyed upon by a whole range of raptors, including buzzards, black kites, falcons and short-eared owls.

Insects

Turning and aeration of the savannah soil is carried out by a variety of insect species such as the dung-beetle *(right)*, which rolls the faeces of herbivores into small balls and buries each in the ground after an egg has been laid in it.

Cleaning-up in the Savannah

Quite a few animals depend upon the successful activities of savannah predators. Big cats only eat the softer parts of their prey after a kill, particularly the liver, which is rich in vitamin A which they cannot synthesise themselves. The remains are taken over by jackals *(below)* and by hyaenas *(above)*, which have powerful jaws and teeth so that they can even break up the bones of carrion. Next in line come the vultures, often joined by marabou storks *(right)* and pied crows. The few remaining fragments are then demolished by hordes of insects, including various ant, fly and beetle species. As a result, all that is left a few hours later is (at the most) a heap of completely cleaned bones.

Locusts

Shortly after mating, the female locust makes a hole in the ground with her ovipositor. After first depositing a layer of viscous mucus, she lays about two dozen eggs. Once the eggs have been laid, the adults die.

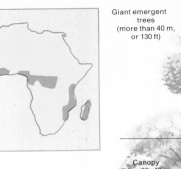

Giant emergent trees (more than 40 m, or 130 ft)

Canopy (from 30–40 m, or 100–130 ft)

Understorey (up to 15 m, or 50 ft)

Extremely sparse undergrowth

Forest Layers
Heavy, continuous rainfall and high temperatures permit the plant species in tropical forests to grow almost without interruption, and the vegetation is also extremely varied. Most of the plants are woody species with tall trunks and, because their foliage develops at a number of different levels, they combine to produce characteristic layers in the forest. By contrast, there are few plant species at ground level since sunlight barely penetrates through the dense tangle formed by tree foliage and epiphytes. Only a few shrubs are found here and there. As a result of the limited availability of light, most trees have tall, straight trunks which are more or less devoid of side-branches until the spreading crown is reached.

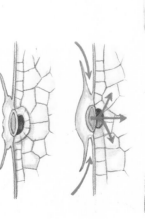

Tropical Fruits
The banana palm *(left)* is a plant with a very large rhizome and a false trunk, sometimes reaching 6 metres (20 feet) in height, formed by a series of interlocking leaf bases. The mango-tree *(right)* produces an amber-coloured resin.

A Convenient Support
Epiphytes are autotrophic (independently nourished) plants which grow on other plants, usually with a long stem or tall trunk. They are not parasites, but merely use these other plants as supports. Typical epiphytic plants are mosses, lichens, orchids and tropical bromeliads.

Strangling Plants
Numerous tropical plants, including various liane species and certain fig species *(right)*, wrap themselves around trees and consequently limit tree growth. In the long term, this intimate relationship can threaten the tree concerned and in some cases even result in its death.

The Equatorial Rainforest

The equatorial forest zone of Africa is exceeded in extent only by the rainforest of Amazonia. In places where the savannah gives way to forest, there is usually a band of bushes, shrubs and green plants which provide a protective screen against fires. As one moves away from the savannah into the forest, the vegetation becomes progressively denser. In the middle of the rainforest, little light can penetrate and the climate is both uniform and humid. In places, swamps are formed. The undergrowth consists of a tangle of ferns, bushes and climbing plants. Any leaves, fruits or branches which fall to the ground rapidly decay and provide food for pig species, porcupines, termites, fungi and bacteria. There are virtually no plants at ground level and, as a result, the only large herbivores present are elephants, buffalo, antelopes and duikers, which feed on leaves and shrubs. Almost all the food available is on the trees and most of the animals are therefore adapted for arboreal life.

Nevertheless, there are a number of species which live at ground level and nest among the bushes, such as guinea-fowl, rails and Congo pea-fowl. When night falls, thousands of bats take to the air in search of food and they in turn are threatened by a deadly predator, the bat hawk *(Machaeramphus)*. A number of insectivorous bird species, such as fly-catchers, shrikes and thrushes, can commonly be found following columns of army ants which put to flight many insects in their path. These voracious ants attack anything which lies in their way, including quite large animals. Even elephants will move away from a column of army ants, since their extremely sensitive trunks might otherwise be bitten.

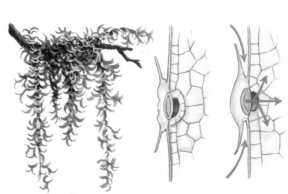

In the Heart of the Forest

Male elephants, which tend to be somewhat irritable at the best of times, will fight relentlessly with rivals when breeding. These large pachyderms lead a nomadic existence and live both in the densest forest and out in the savannah. Other animals, by contrast, never leave the forest; these include pangolins, various antelope species, wild pigs, wild boar, okapis *(below right)* and large-bodied bats which can find all they need to eat in the trees. The trees are also inhabited by a large number of snakes which prey on eggs, nestlings and adult birds, while themselves falling prey to a variety of raptors such as the Congo serpent eagle which has particularly large eyes and can therefore spot its prey more easily in the semi-obscurity of the forest.

The Green Mamba

The green mamba is noteworthy among poisonous snakes in that its venom takes effect very rapidly.

African Apes and Monkeys

Social groups of gorillas *(above right)* have particularly large home ranges, within which they forage for shoots and leaves. Each group typically has one large-bodied male with grey hairs on his back (silver-back male). Chimpanzees *(below)* can climb around in trees with considerable agility and they live in feeding groups of flexible composition which move around very actively in search of food, especially fruits. Among the leafy branches of the large trees a number of monkey species can be found, including both cercopithecines (guenons) and colobines (leaf-monkeys). Mandrills *(above left)*, like their close relatives the savannah baboons, form large social groups which spend most of their time on the ground.

Predators

The most treacherous enemy of monkeys and antelopes in the African forest is the leopard *(above)*, a magnificent big cat whose great agility allows it to pursue its prey through the trees. The rainforest is also inhabited by a variety of wild cats, mustelids and viverrids. The genet *(right)* is one of the most graceful representatives of the family Viverridae (civet family). These relatively small carnivores are very agile and at the same time very cautious, hunting principally by night and concentrating on rodents and reptiles.

Nectar-feeding Bats

African long-tongued fruit bats *(Megaglossus)*, as their name implies, have specially adapted tongues for feeding on the nectar of flowers in the rainforest.

55

Mountains and High Plateaux

With the exception of the Atlas mountains, Africa lacks any prominent mountain-chains. Instead, there are high plateaux with isolated mountain peaks, notably in East Africa. The Ethiopian high plateau, which is broken up by numerous precipices and ravines, bears witness to a previous age when water was present in abundance; but it is now surrounded by deserts and semi-arid lands. The high Simien massif forms a vast high plateau, with a rich plant community, which is the meeting-ground for European, Ethiopian and African faunas. Large herds of gelada baboons live here in addition to a multitude of rodents, such as swamp rats *(Otomys)*, which attract flocks of predators. The highest peak of Ethiopia, the Rasdajan (4,620 metres or 15,250 feet), is found in the Simien massif, the kingdom of black vultures, eagles and pied crows. The high plateaux of East Africa are covered with grassy plains lying at an altitude of 1,200 to 2,000 metres (4,000 to 6,600 feet). Here are found the highest peaks of the African continent, usually of volcanic origin, including Kilimanjaro (5,790 metres or 19,340 feet) and Mount Kenya (5,199 metres or 17,058 feet). These two peaks, along with the Ruwenzori range, are the only ones to have a permanent snow-cap. In the luxuriant forests surrounding these peaks live many large-bodied mammals, such as elephants, rhinoceroses, giant forest hogs, buffalo, and (in some areas) mountain gorillas, ibex, hyraxes and the very occasional leopard.

Stratified Vegetation Zones

As altitude increases, there is a succession of plant formations. There is a lower zone of *Podocarpus* conifers and cedars, occupied by large mammals such as gorillas and rhinoceros. Then comes a montane bamboo zone which provides food for birds and insect-eating monkeys and which is sometimes visited by elephants. Further up still there are clearings carpeted with lady's-mantle where buffalo and antelope live. After this, there are tree-heaths, then tree senecios and, finally, high-altitude heathland.

Kilimanjaro
The gigantic volcanic dome of Kilimanjaro, the highest peak in Africa, dominates the high plateaux of Kenya and Tanzania. The principal crater, Kibo, is surrounded by an ice cap, from which flanges spread out radially down the slopes.

5,800 m (19,000 ft)

Lobelia

Swamp rat

Tree senecio

4,800 m (15,700 ft)

Hyrax

Three-heath

Leopard

Bamboo

Bushbuck

3,000 m (9,800 ft)

Guenon

2,400 m (7,900 ft)

Thick-tailed bushbaby

Bongo

Podocarpus

Mountain gorilla

1,500 m (4,900 ft)

Rhinoceros

Elephant

Acacia

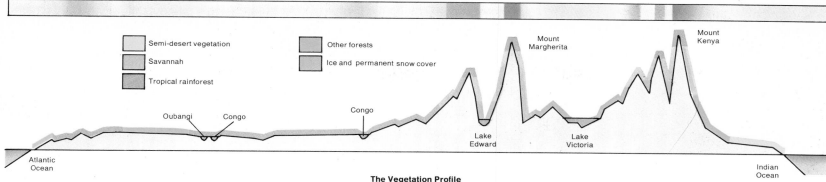

The Vegetation Profile

There are a number of factors which limit plant growth and hence influence the geographical distribution of plant species: level and annual distribution of rainfall; soil type; exposure to sunlight; temperature; and so on. Collectively, these are referred to as climatic and physical factors. The diagram *above* shows the different plant associations which occur in sequence from the Atlantic coastline to the Indian Ocean coastline of Africa. It can be seen that the presence of bodies of water, even far inland, gives rise to humid local microclimatic conditions with resulting luxuriant plant growth. The further one moves away from the great oceans or from the lakes, the more arid the conditions become, culminating in semi-desert and desert landscapes.

Pink Flamingos

Progressive loss of water from certain lakes in the Rift Valley has led to gradual increase in the salinity of the water, and this has generated an ideal habitat for certain micro-organisms, such as blue-green algae and diatoms. These organisms provide food for thousands of pink flamingos which gather on the shores of these lakes. The flamingos have a specially structured beak which permits them to filter tiny organisms from the water.

Mount Kenya

As is typical of the high African mountains, Mount Kenya *(above)* was produced by volcanic activity which took place during the Tertiary and Quaternary eras in East Africa. Mount Kenya consists of a core of crystalline rock forming a truncated cone rising up above a foundation of volcanic rock.

Tree-Heaths

African tree-heaths, belonging to the family Ericaceae, can grow to a height of 18 metres (60 feet). Heathland presents ideal feeding conditions for many animal species because of the availability of fine grasses and succulent plants.

The Rift Valley

The map *(right)* illustrates the general contours of the Rift Valley, a gigantic East African tectonic trough. This trough includes the Red Sea and the great African lakes; to the south it is subdivided. The Rift Valley was created during the Tertiary era through tensions generated by the slow movements of the tectonic plates constituting the Earth's crust. This process is still continuing today.

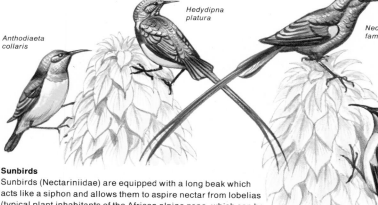

Anthodiaeta collaris
Hedydipna platura
Nectarina famosa
Anthrepetes longuemarei

Sunbirds

Sunbirds (Nectariniidae) are equipped with a long beak which acts like a siphon and allows them to aspire nectar from lobelias (typical plant inhabitants of the African alpine zone, which can be found growing up to altitudes of 4,000 m, or 13,000 ft). Through their nectar-feeding habits, these birds contribute to the pollination of the lobelias.

The Hyrax

Among the giant lobelias of Kenya, at an altitude exceeding 4,000 metres (13,000 feet), lives the rock hyrax *(Procavia)*, a diminutive relative of the elephant which is preyed upon by leopards.

Mountain-living Monkeys

The small green monkey *(right)* lives on the edges of forests and in relatively open woodland on mountain slopes up to altitudes exceeding 3,000 metres (10,000 feet). These monkeys feed mainly on the ground during the daytime, but they never move far away from the trees, which they use for sleeping at night and as temporary refuges if danger threatens during the day. These lively monkeys live in groups of twenty to thirty containing several adult males in addition to females and offspring.

Paradise widow-bird *(Steganura paradisaea)*
Melba finch *(Pytilia melba)*

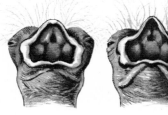

A Parasitic Relationship

Widow-birds (or whydahs) are a specialized subfamily of African weaver-birds which have developed a special parasitic relationship with other birds, the waxbills (family Estrildidae). In contrast to the European cuckoos, where the single parasitic offspring is reared alone by the adoptive parents, the offspring of the African widow-bird is reared along with the host's own offspring. Each widow-bird species parasitizes the nests of a particular waxbill species, and its eggs are closely matched in form and colour (a case of "mimicry"). The widow-bird nestling also resembles the offspring of the host in that the same pattern of coloured spots is present in the gape to elicit feeding.

57

Large Rivers and Marshland

The Nile, the greatest river in the world, has many tributaries which are themselves fed by numerous lakes, including the great Lake Victoria. Along the banks of the Nile, enormous marshland areas are produced. The largest of these marshland areas is the Sudd in southern Sudan. In such areas, the vegetation is dominated by papyrus, reeds and various aquatic plant species. Sometimes, the water has a low oxygen content because of the mass of vegetation, and this obviously poses serious problems for the plants and animals living in this habitat. One marsh-living worm has produced a partial solution to this problem, since it behaves like the earthworm and facilitates both aeration and decomposition of plant debris. A wide variety of animal species can be found in the marsh environment. Among others, there are sitatungas (large-bodied antelopes), hippopotamus, crocodiles, monitor lizards and several aquatic bird species, including herons, pelicans, shoebill storks and hammer-headed storks.

The Great African Rivers
During the rainy season, many of the large African rivers, such as the Niger, the Senegal, the Volta and the Nile, overflow and produce large expanses of marshland known as "floodplains". These zones attract a large number of migratory, aquatic bird species. For instance, the ducks which leave Eurasian taiga and tundra zones as winter approaches migrate down to the marshes formed by the River Niger. *Above:* The Zambesi, a large river in southern Africa which feeds the famous Victoria and Livingstone Falls. *Right:* Marshland in Ghana.

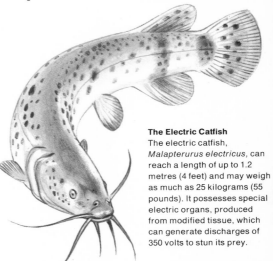

The Reptiles
Crocodiles *(above)* contribute to the maintenance of equilibrium in the fish populations inhabiting rivers and marshes. Monitor lizards *(right)* are carnivorous and include both eggs and nestlings in their diet.

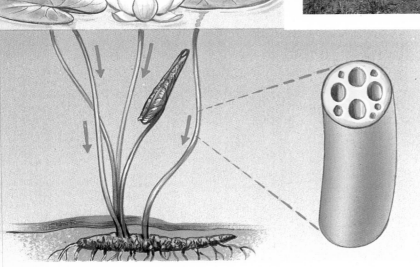

Respiration in Aquatic Plants
When they are dilated, plant cells take on a rounded form and intercellular spaces are formed, as shown in the enlarged illustration *(left)*. These air-filled spaces communicate with one another and permit a flow of oxygen to the plant cells. In fact, plants do not have a gas transport system comparable to that found in many animals. In aquatic plants, the air-filled spaces in the cells are particularly well-developed to offset the particular difficulties which these species face.

The Electric Catfish
The electric catfish, *Malapterurus electricus,* can reach a length of up to 1.2 metres (4 feet) and may weigh as much as 25 kilograms (55 pounds). It possesses special electric organs, produced from modified tissue, which can generate discharges of 350 volts to stun its prey.

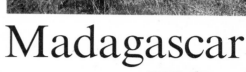

Madagascar

The Hippopotamus
Despite the squat shape of its body, the hippopotamus is an agile swimmer. This mammal can remain submerged for hours at a time, with only its nostrils and its eyes protruding above the water surface. The hippopotamus plays a vital role in the maintenance of the nutrient cycle of rivers and lakes. Its dung provides a large amount of fertilizer which encourages the development of tiny blue-green algae—an essential food for numerous fish, including the perch-like *Tilapia* which is a major part of the diet of local human inhabitants. The hippopotamus also ensures that water continues to circulate, by maintaining passages through the tangle of papyrus and floating vegetation as it moves around at night to feed on the vegetation on meadows lining the banks.

The Shoebill Stork
The shoebill stork *(right)* lives among the papyrus and seeks out frogs and fish while striding up and down over the muddy substrate. The jacana *(left)* has very long toes which permit it to walk across the floating leaves of marsh-living plants.

Turtles
The flourishing vegetation of tropical rivers and swamps provides a very suitable habitat for numerous turtle species. Soft-shelled turtles of the genus *Trionyx (above)* have a flat carapace devoid of scales, a feature which distinguishes them from other turtle species. They keep to the bottom of rivers and lakes, where they hunt for fish and molluscs.

The Wily Fishers
Along the banks of lakes and rivers can be found large flocks of pelicans which spend the day hunting for fish. To do this, they make use of a very effective technique. They swim along in a tight-packed line, thus obliging the fish to stay close to the bank; then they all plunge their beaks into the water at the same time to catch a large number of prey in one fell swoop.

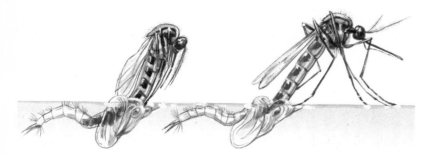

The Geographical Distribution of Malaria
Malaria is primarily found in the warmer regions of the world, since the protozoan responsible for the disease *(Plasmodium)* will only develop in the bodies of mosquitoes at temperatures exceeding 24 °C.

Mosquitoes and Malaria
The carriers of malaria are mosquitoes of the genus *Anopheles*. These mosquitoes lay tiny, spindle-shaped eggs which are each equipped with a small air-filled cavity which enables them to float. The larvae, after hatching, swim at the surface of the water, filtering out pollen, bacteria and spores as their food. If danger threatens, the larvae take refuge at the bottom; but they are obliged to return to the surface regularly in order to breathe. The diagram *above* shows the final phases of the emergence of the adult insect. Unfortunately, these mosquitoes are still extremely common in humid areas of the tropics.

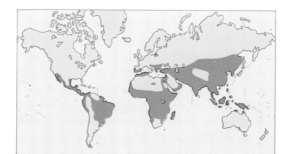

Before the advent of the Secondary era, this huge island in the Indian Ocean was combined with Africa, Antarctica, India, Australia and South America in the gigantic supercontinent of Gondwanaland. The fragmentation of this supercontinent into separate blocks which subsequently drifted apart gradually gave rise to the present distribution of land-masses in the southern hemisphere.

The Vegetation
Three major vegetation zones can be distinguished in Madagascar: the east coast tropical rainforest, with evergreen species such as screw-pines (*Pandanus utilis*) and raffia palms; the savannah on the high central plateau; and the xerophytic vegetation adapted for the semi-arid conditions prevailing in south-west Madagascar, including various spiny plant species and the thick-set euphorbias (spurges). The flora shows little affinity with that of Africa and instead shows some surprising resemblances to the plants of South Asia, Indonesia and Australia. It seems likely that there was once a basic southern flora which became differentiated later after the break-up of Gondwanaland.

Pachypodium

Aloe   Baobab

The Animal Species
Numerous animal species are found only in Madagascar, such as the lemurs *(below right)*, the tenrecs (hedgehog-like tailless mammals), a large number of chameleons, large-bodied tortoises and a number of special carnivores such as the fossa. *Below left:* The vasa parrot.

BAFFIN BAY
BANKS ISLAND
BEAUFORT SEA
VICTORIA ISLAND
GREENLAND
BROOKS RANGE
BAFFIN ISLAND
Davis Strait
Great Bear Lake
ALASKA
Yukon
ALASKAN RANGE
MACKENZIE MOUNTAINS
Mackenzie
Great Slave Lake
HUDSON BAY
ALEUTIAN ISLANDS
ROCKY
COAST
LABRADOR
VANCOUVER ISLAND
CANADIAN PRAIRIES
Lake Winnipeg
Missouri
MOUNTAINS
RANGE
Great Lakes
Saint Lawrence
NEWFOUNDLAND
GREAT
PLAINS
Ohio
APPALACHIAN MTS
ATLANTIC
COLORADO DESERT
Mississippi
OCEAN
PACIFIC
Rio Grande
CALIFORNIAN PENINSULA
SIERRA MADRE
GULF OF MEXICO
HAWAIIAN ISLANDS
OCEAN
CARIBBEAN SEA

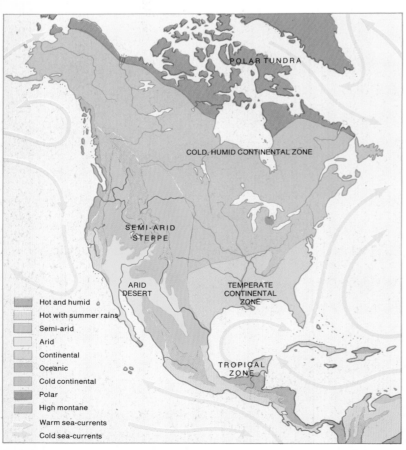

POLAR TUNDRA

COLD, HUMID CONTINENTAL ZONE

SEMI-ARID STEPPE

ARID DESERT

TEMPERATE CONTINENTAL ZONE

TROPICAL ZONE

Hot and humid
Hot with summer rains
Semi-arid
Arid
Continental
Oceanic
Cold continental
Polar
High montane
Warm sea-currents
Cold sea-currents

NORTH AMERICA

North America offers a large number of different natural habitats, ranging from the vast Arctic zone, whose shores are invaded by the ice of the polar ice-pack, to the arid desert conditions of the Great Basin or Colorado Desert in southern California.

South of the Arctic tundra zone there is an immense expanse of boreal conifer forest which is extremely uniform in its composition and interrupted only by a latticework of lakes of all shapes and sizes. The particular arrangement of the mountain-chains, which block off humid winds from the ocean, accounts for the existence of a vast prairie in the heart of the great continent. Of course, this region has now been largely turned over to cultivation and has therefore lost its original distinctive character, but just a century ago immense herds of bison roamed over the plains. To the north, the prairie gives way to the Piedmont forest, a deciduous forest area which is the last testimony to the original huge expanse of tree cover, and to the depression containing the great lakes. To the south, the prairie also becomes more and more occupied by

Contrasting Landscapes of North America
The photograph *(above)* provides a view of St. Mary Lake in Montana. The conifer forests, lakes and mountainous peaks with permanent snow cover on their slopes, combine to produce an alpine type of environment which is characteristic of northern areas of the United States. The other photograph *(right)* shows a spectacular example of erosion in Utah. Agents of erosion, beginning with running water and ending with the corrosive action of winds, have clawed away the soft clay to leave a bizarre landscape marked by multicoloured pinnacles.

trees. Intensive human activities have left their traces everywhere. This has, indeed, given rise to an extremely serious problem—pollution. When travelling across the United States, one can find indestructible plastic containers and all kinds of other packaging materials in places which otherwise seem to be uncontaminated, for example on mountain peaks, in the centre of a desert or in the meandering channels of a foetid swamp. But the record for pollution levels is held by the beaches. Fortunately, the United States of America was also the first country to establish the concept of national parks and thus to provide a restraint on the destruction of natural environments. The first national park, Yellowstone, was created in 1872 and was soon joined by others, such as those of the Grand Canyon, the Sequoia-Kings Canyon, the Everglades and Death Valley. There are now two hundred and eighty protected areas in the United States and some of them, such as the coastal forests of the Pacific and the swamps of Florida provide a vast subject of study for naturalists, with many secrets still to be discovered.

The Great North and the Taiga

The northernmost vegetation zone of the American continent (the tundra), just beneath the Arctic polar circle, accounts for one fifth of the total land area. This region, which is covered in ice in winter, is carpeted with a flourishing vegetation composed of dwarf plant species—mosses, lichens and a profusion of flowers which attract silkworms and other moths and butterflies. Because of the presence of permafrost, the soil is saturated with water in summer and, as a result, the tundra is infested with mosquitoes which both in number and in vexatiousness provide more than a match for their tropical counterparts. The large numbers of insects present make this region a favoured haunt for many bird species which make their nests in the tundra. But the basic foodstuff is constituted mainly by lichens, the staple diet of such large mammals as the caribou and of small rodents such as lemmings. South of the tundra lies an equally vast zone of very uniform appearance, the boreal conifer forest, with a network of lakes of all sizes, pools and huge expanses of marshland produced by melting ice. Because of the influence of warm air-currents of oceanic origin, conifer forests are also found along the Pacific coastline of Alaska, of British Columbia and of the Yukon, and this tree cover extends into inland valleys. This forest zone incorporates three types of vegetation: ferns and bushes; small conifers and broad-leaved tree species; and gigantic conifers such as the Douglas fir, which may reach a height of as much as 100 metres (330 feet).

A Landscape of Glacial Origin in the Northern United States
The typical U-shaped cross-section of St. Mary's Valley in the Glacier National Park (Montana) bears witness to the predominant influence of glacial erosion in shaping the landscapes of this area. During the Quaternary era, the North American continent was invaded by ice at the time when glaciations were taking place in Europe. In North America, the ice extended down to the 40th parallel, forming enormous ice-sheets thousands of metres thick and sweeping down either from the coastal Rocky Mountain chain or from Hudson's Bay. The latter, known as the Laurentide ice-sheet covered almost the entire area of Canada and fused in the north with the Greenland ice-sheet, which is still present today.

Permafrost
The Canadian high plateau and Alaska are characterized by harsh low temperature conditions throughout the year. During the hottest months, the temperature never exceeds 10 °C and in winter average temperatures range from —22 °C along the coast to —32 °C inland. The soil is almost permanently frozen (permafrost). In summer, there is some thawing but this only affects a superficial layer of the soil, the mollisol, which is no more than 10 centimetres (4 inches) deep.

Plant Associations
The diagram *(right)* shows the succession of plant associations which is found, progressing from a marshy environment to a dry land area with acid soil, where conifers grow. The reproductive cycle of conifers is illustrated in the circle *(above right)*. Microspores produced by male cones are carried by the wind and eventually reach the ovules contained in female cones, so that fertilization can take place. The seed which is thus produced falls to the ground and produces the new plant.

A) Fertilization
B) Seed-formation
C) Seed dispersal
D) Germination and development of the new plant

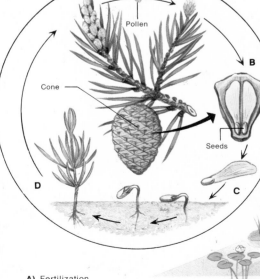

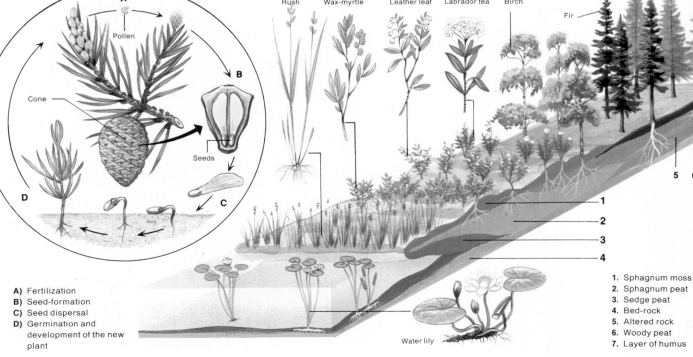

Rush   Wax-myrtle   Leather leaf   Labrador tea   Birch   Spruce   Fir

Pollen
Cone
Seeds

Water lily

1. Sphagnum moss
2. Sphagnum peat
3. Sedge peat
4. Bed-rock
5. Altered rock
6. Woody peat
7. Layer of humus

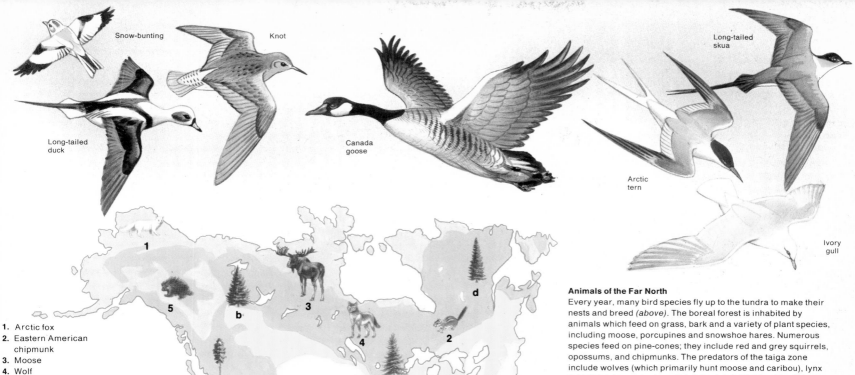

Snow-bunting

Knot

Long-tailed skua

Long-tailed duck

Canada goose

Arctic tern

Ivory gull

1. Arctic fox
2. Eastern American chipmunk
3. Moose
4. Wolf
5. Porcupine
a) Pine-tree *(Pinus contortus)*
b) Balsam fir *(Abies balsamica)*
c) Larch *(Larix laricina)*
d) Spruce *(Picea glauca)*

Tundra   Conifer forests   Semi-deciduous forests

Animals of the Far North

Every year, many bird species fly up to the tundra to make their nests and breed *(above)*. The boreal forest is inhabited by animals which feed on grass, bark and a variety of plant species, including moose, porcupines and snowshoe hares. Numerous species feed on pine-cones; they include red and grey squirrels, opossums, and chipmunks. The predators of the taiga zone include wolves (which primarily hunt moose and caribou), lynx (which feed on rodents and birds), mink (which hunt for fish, reptiles and birds), martens (which are deadly enemies of squirrels), and various birds of prey, such as the bald eagle, the osprey and the Virginian eagle-owl.

Relationships between Species

Man is able to benefit from the continuing struggle for survival by exploiting the antagonism which exists between organisms living on the same plant species, such as the caterpillar of the procession moth and the red ant. The presence of a natural predator on a parasite which one wishes to eliminate creates a "balance of force" between the two antagonists, thus limiting the damage done to plants. The procession moth caterpillar lives on pine-trees and voraciously devours the pine-needles. The illustration *(right)* shows how to combat this parasite without resorting to chemical agents, by encouraging the spread of a particular red ant species which is a natural predator on the caterpillars.

The Taiga or Boreal Forest

The great American boreal forest is an extremely uniform habitat interrupted only by a multitude of lakes and marshes of various sizes. *Above:* A view of the American taiga with the Rocky Mountains in the background.

A Bark-eater
The North American porcupine *(Erethizon)* feeds on the bark of trees *(right)* and damages growing conifers. The second photograph *(far right)* shows a Canadian capercaillie.

A Great Acrobat
Thanks to the membrane stretched between its limbs on each side of the body, the flying squirrel *(Glaucomys)* is able to glide about high up in the trees. The illustration *(right)* shows various phases of landing. This graceful animal has a very varied diet compared with that of other squirrels, since it feeds on eggs and nestlings as well as on pine-cones.

The Coastal Forest
The dense coastal forest bordering the northern Pacific is inhabited by a great variety of animal species, though birds predominate: ducks, geese, loons, cormorants, ptarmigans, crows, eagles, falcons and (in summer) humming-birds. There are several herbivorous mammals, including Dall sheep, North American mountain goats and mule-deer. Numerous predators feed on the many herbivorous species present, notably foxes, wolverines, martens, mink and lynx. The largest carnivore in this forest region is the brown bear *(right)*, which has diversified into a number of subspecies.

63

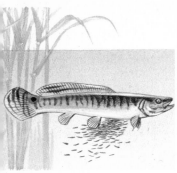

Freshwater Animals

The bowfin (Amia calva) is characterized by a long dorsal fin which lacks spines. This fish has an unusual breeding pattern. The male prepares a nest with aquatic plants for the reception of the eggs and he guards them during their development. After hatching, he continues to watch over the young fish. The freshwater habitats of this region are typically inhabited by perch, trout, salamanders, toads, frogs, turtles, crayfish, bivalve molluscs and the larvae of numerous insect species. There are also a large number of mammals which obtain their food from the water: muskrats, water-voles, otters and mink. The habitual bird inhabitants include kingfishers, ducks, geese, swans, loons and ospreys.

The Great Lakes

Most of the lakes of North America, ranging from the smaller ones such as Wallenpaupack Lake (above) to the Great Lakes lying on the frontier between Canada and the United States, were apparently formed subsequent to the Quaternary glaciations. The enormous basins of the Great Lakes were formed through a combination of river erosion during the Mesozoic era and glacial erosion during the Pleistocene (the first period of the Quaternary era). In fact, Lake Superior, with a surface area exceeding 84,000 square kilometres (33,000 square miles) is the largest lake region in the world, if the Caspian Sea is excluded. Lake Superior lies in an excavation of extremely ancient metamorphic rocks belonging to the Canadian continental block. The other four lakes lie in a zone of softer Palaeozoic rocks. The gigantic Pleistocene ice-cap which covered this area during the various phases of glaciation considerably modified the pre-existing landscape, which contained a network of rivers. The last glacial expansion, which terminated about 14,000 years ago, was characterized by vast fingers of ice which deposited a concentric pattern of moraines.

Salmon

After they have remained in the sea for a period which ranges from one to several years, according to the species, salmon (above) undertake a long migration back to the river where they were born. Swimming upstream, they progress up rapids and waterfalls with powerful leaps until they reach the breeding-grounds.

The Great Lakes and the Appalachians

In the central region of North America, south of the taiga zone, three different areas can be distinguished: the depression occupied by the Great Lakes; the broad-leaved forests of Piedmont; and parklands, which are transitional between prairie and deciduous forests, with scattered large trees and widespread bushes. From the point of view of human habitation, this is one of the most developed regions and has therefore been subject to considerable modification. Only a few vestiges remain of the immense expanse of forest which was once present. Nevertheless, the Great Lakes and the rivers have retained a healthy fish population. In addition to common species such as trout, bass, pike and perch, there are a number of unusual fish including the paddle-fish (which has a spoon-shaped mouth). The Appalachian region has also been greatly modified by human activity. During the great colonization period, large areas were deforested either to create new pastureland or to develop iron mining. The region extends for more than 2,000 kilometres (1,250 miles) from New Brunswick to north-west Georgia, and its average width is about 300 kilometres (185 miles). In the first third of the region, lying to the north and stretching from Fundy Bay to the mouth of the Hudson River, the mountains sweep down to the coast. To the south, however, the Appalachian mountain-chain swings inland and a vast coastal plain is present to the east. The Appalachians actually incorporate a number of mountain-chains, which are arranged in sequence to the north and are virtually parallel in the south. The Maine region is characterized by a forest of boreal type in-

Forming a Circle against Predators

During the night, bobwhites settle down in a circle. This behaviour has a double function: firstly, it provides protection against the cold, especially in autumn and winter; secondly, if a predator approaches the birds all take off rapidly in different directions and thus confuse the potential aggressor.

THE SAINT LAWRENCE SEA-WAY

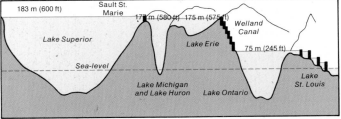

habited by animal species which are typically associated with conifer forests.

The undergrowth is of particular interest because of the large number of fern and moss species and especially because of the incredible diversity of fungi which grow almost everywhere. These fungi exhibit a remarkable variety of form and colour. There is even one species which is green, an extremely unusual colour for a fungus since fungi do not produce chlorophyll.

The southern mountain-chains of the Appalachians are largely covered with deciduous tree species with foliage extending up to a height of 30 metres (100 feet): beech-trees, lime-trees, walnut-trees, magnolias, maples, oaks and poplars are the dominant species. In this region, the vegetation is stratified and in summer the crowns of the highest trees form a dense canopy which cuts off much of the light from the undergrowth layer. Although many original animal inhabitants of these forests have become rare or have disappeared altogether (such as the wolf and the mountain lion), there is still a varied fauna. Huge herds of deer find food and refuge in the heart of the forest, while wild rabbits, marmots, moles, shrews, grey squirrels and chipmunks are found everywhere, along with numerous bird species. Along the coastal fringe, which is flat but very irregular in outline, the tide limit is extremely variable and there are large sand-banks which shift continuously through the action of storms. In the swamps and lagoons, which are very common, there is a characteristic swampland vegetation and some growth of bushes. There are also numerous water-courses which irrigate the coastal plain east of the Appalachians.

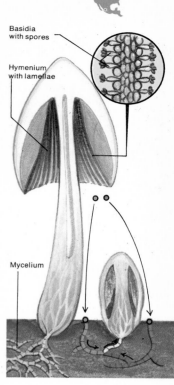

The Life-Cycle of the Mushroom
In the reproductive cycle of the higher fungi (Eumycetes), there are two morphologically different generations. The basidia *(see diagram above)* produce two types of spore (positive and negative) which each produce thread-like mycelia with a single nucleus and with unlimited growth potential. When filaments from positive and negative spores (i.e. of different sexes) meet, the cells fuse and produce sporophytes with two haploid nuclei. Multiplication of the sporophytes leads to the formation of a fruiting body, the mushroom. Fusion of the two nuclei to produce a single diploid nucleus takes place in the basidia, which are special organs located beneath the cap of the mushroom. The life-cycle is completed when diploid cells form haploid spores by meiosis and the spores subsequently ripen and fall to the ground.

The Appalachian Region
Carnivores abound in the luxuriant deciduous forest growth of the Southern Appalachians: bobcats, raccoons *(left)*, weasels and skunks are particularly characteristic. In certain areas there are also black bears, which are primarily plant-eaters. Birds are found everywhere and will not hesitate to approach human settlements: gulls, crows, pigeons, wild duck and falcons. Tortoises and snakes, particularly vipers and rattlesnakes, are very common. Among the numerous insect species, the solitary wasp deserves a special mention. This wasp bores circular holes in trees and builds thimble-shaped nest-chambers with leaf fragments. Each such nest-chamber contains a single egg and food reserves for the future larva.

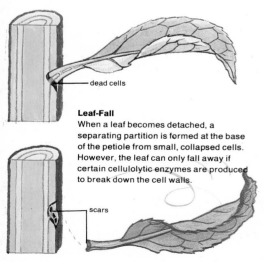

Leaf-Fall
When a leaf becomes detached, a separating partition is formed at the base of the petiole from small, collapsed cells. However, the leaf can only fall away if certain cellulolytic enzymes are produced to break down the cell walls.

Glaciations
Zones affected by glaciations are characterized by particular deposits from glaciers and rivers, such as "eskers" and "kame terraces". The upper diagram *(right)* shows a glacial region with subglacial streams and marginal water-courses. When the ice melts, the sediments deposited on the edges form kame terraces, while eskers are formed as narrow ridges through the filling of subglacial canals.

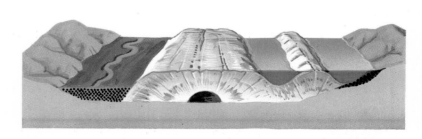

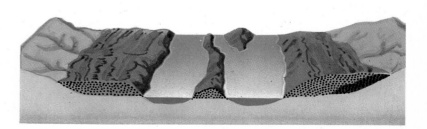

The Prairie
The prairie is a plant association whose existence depends upon limited rainfall, (such that the growth of trees is prevented) and winter snowfall (which protects the soil from freezing). It is composed of annual and perennial plant species characteristic of regions south of the boreal forest, with grasses predominating. Blue grama and buffalo grass constitute 70 to 90 percent of the plant cover. *Extreme left:* A herd of bison; *Left:* A prairie dog.

The Great Prairie

The heart of the North American continent is constituted by an immense prairie zone, which is nowadays largely under cultivation. This prairie zone has a relatively low rainfall because the oceans are far away and mountain-chains block off humid winds. It can be said that the main architects of the prairie have been two animals, the prairie dog and the bison. Prairie dogs are now limited to a few restricted areas, but originally there were millions of them throughout the region, constructing their communal burrow systems with countless underground passages. In past centuries, these "subterranean towns" were inhabited by some three hundred million families of prairie dogs. Such large numbers of animals obviously displaced vast amounts of earth, loosening and aerating the soil and permitting water to penetrate deep down. In addition, the prairie dogs

kept the grass regularly cropped and thus stimulated its growth. Their activities in this direction were augmented by those of the enormous herds of bison which wandered across the entire American prairie zone, also numbering in their millions. Unfortunately, human intervention brought the beneficial activities of these two animals to an abrupt halt. Prairie dogs have been systematically wiped out, because their burrow systems made the ground unstable, with a consequent risk of injury for horses. Similarly, the bison were practically exterminated at the time of construction of the railways. In the wake of the disappearance of these original animal inhabitants, the prairie gradually changed in character and vast areas have now been invaded by herbaceous plants and bushes.

Predators of the Prairie
Small burrowing mammals such as prairie dogs are preyed upon by a variety of predators. Rattlesnakes pursue them down into their burrows, while prairie falcons *(above)* swoop down upon them when they venture out to feed. The badger, itself a large-bodied and robust burrowing mammal, digs up burrow systems and eats any animals it meets on the way. Another dangerous predator of prairie dogs is the blackfooted ferret, which has a slender, agile body and can pursue its prey through the narrowest of underground passages. *Below:* A coyote, an animal of wide dietary habits found both in forests and in semi-arid habitats.

Burrowing Animals
In addition to prairie dogs, which are in fact relatives of squirrels, there are numerous animals which take refuge below ground. The burrowing owl *(left)* makes use of the burrows of the prairie dogs and lives in peaceful association with them, apart from occasionally devouring their young. This bird of prey usually feeds on a variety of rat species which inhabit the prairie. *Below:* A number of other prairie inhabitants: **1.** Woodchuck; **2.** Pocket gopher; **3.** Blind-worm; **4.** Lemming; **5.** Pocket mouse, showing its very elaborate burrow (**a** = living chamber; **b** = food store; **c** = access tunnel; **d** = escape tunnel).

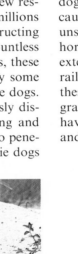

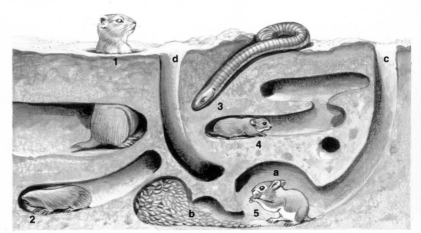

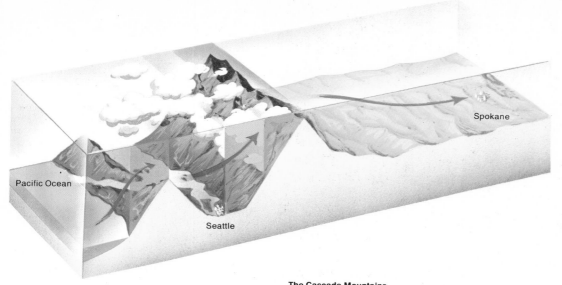

Rattlesnake

The Pronghorn Antelope
Just like the bison, the American pronghorn antelope *(above)* almost became extinct because of indiscriminate hunting. The male of this species, which is similar in size to the roe-deer, has permanent bony horns covered with a keratinous envelope which is renewed each year. *Right:* An adult skunk with its offspring.

The Cascade Mountains
This mountain-chain is one of the large series of chains which run along the west coast of North America. Here, however, the peninsula of Mount Olympus constitutes a major factor in climatic differentiation. The winds which arrive loaded with humidity absorbed over the Pacific Ocean are forced upwards as they pass over this first mountain-chain. The subsequent fall in altitude, with its accompanying warming effect, permits the air to conserve its humidity, which is then deposited on the western slopes of the Cascades, generating a wet climate (Seattle region). On the eastern flanks of the Cascades, however, the conditions are extremely dry and the area is essentially a desert.

Reptiles
The teeth of reptiles are not used for mastication, for these animals swallow their prey whole. Their teeth merely serve as instruments for prey capture. In the viper family, there are mobile, hollow fangs connected to poison glands. When the mouth is closed, the fangs lie retracted along the palate.

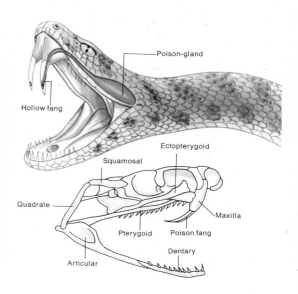

In many reptile species, the upper and lower jaws can move independently, permitting the mouth to be opened wide so that the prey can be swallowed whole.

Insects are Economical with Water
Insects which live in a terrestrial environment must, just like vertebrates, face the problem of dehydration. Aquatic arthropods dispose of metabolic waste in the form of ammonia, whereas insects eliminate their breakdown products in the form of uric acid, which is a less toxic substance requiring little water for its disposal. *Right:* Illustrations of some of the insects found on the prairie.

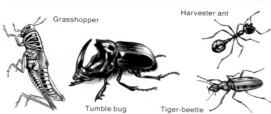

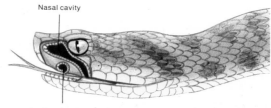

Snakes are equipped with two special sensory systems, the paired Jacobson's organs, which are concerned with smell and taste, and (in some species) special pits anterior to the eyes which are sensitive to heat and allow detection of warm-blooded prey without the need for any other sensory information.

Prairie Birds
In the course of their long voyage from north to south, migratory birds stop over in the prairie to feed on seeds and insects. But this region is also inhabited by a number of gallinaceous bird species which spend most of their time on the ground. Sage grouse *(left)* are distributed throughout the semi-arid prairie areas in the west, which are dominated by *Artemisia* bushes. Wood grouse, by contrast, prefer the high grass region in the east, where they are solitary in habits during the summer and gregarious in winter.

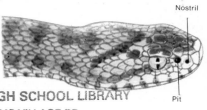

The Mountains and Deserts of the West

Along the Pacific margin of the North American continent run two parallel chains of mountains. The conifer forest found on the northern slopes of the mountain-chain nearest the sea is inhabited by deer, elk and innumerable coyotes. Further south, where the outer chain joins up with the inner chain, broad-leaved deciduous trees are intermingled with the conifers to form one of the most luxuriant forests in the world. The southern region of the forest on the outermost mountain-chain is, in fact, the last refuge of the great sequoia trees. Just behind this part of the outer chain lies the impressive Sierra Nevada, extending over a distance of more than 800 kilometres (500 miles), with its snow-capped peaks towering above a sea of dry grassland with just an occasional bush. It is in this massif that forests of giant sequoias and the famous Kings Canyon are located. The conifer forests of the Rockies constitute an ideal habitat for bears (especially black bears, which are very common) and, at higher altitudes, for a number of hoofed mammals, including bighorn sheep. The northern forest areas are also inhabited by large herds of wapiti.

The Basin Region (Nevada Desert) is the largest

The Rockies
The Rocky Mountains essentially form the backbone of North America, extending over a distance of almost 4,500 kilometres (2,800 miles) from the frozen tundra of Alaska to the scorching sierras of Mexico along the western side of the continent. The Rockies owe their origin to a number of orogenic (mountain-building) events which began in the Palaeozoic era and continued right up to the Caenozoic era. The presence of volcanoes which have remained active until quite recent times, such as Mount Rainier *(above)* in the Cascade mountain range, bears witness to the fact that orogenic processes have not yet come to an end. Nevertheless, the Caenozoic has been a relatively calm period during which processes of erosion and levelling have predominated, culminating in the phase of rejuvenation during the Pliocene which determined the main structural features that we see today. The Rockies are a complex assemblage of several subsidiary mountain-chains, with the highest peaks (exceeding 4,000 metres, or 13,000 feet, in some cases) lying in the south.

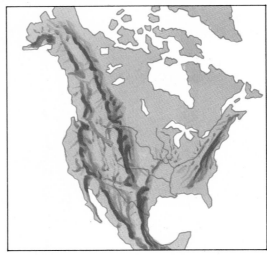

Desert Vegetation
In America, steppe dotted with sagebrush gives way in the south to desert landscapes which stretch from Arizona to Mexico. Where the land is flat, plants of the genus *Yucca (below right)* and cactuses *(above* and *above right)* predominate. In mountainous and rocky areas, on the other hand, agaves (American aloes) are more common. Various prickly-pear cactus species are also characteristic and one particularly noteworthy plant is the saguaro cactus, which can grow as high as 10 to 15 metres (30 to 50 feet).

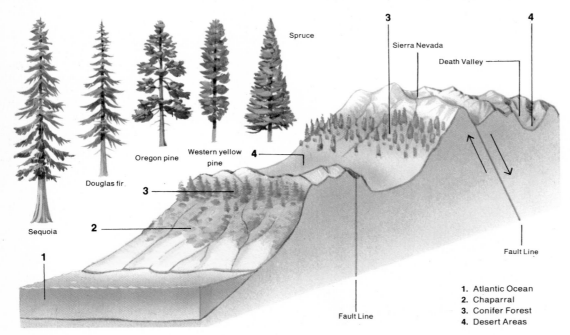

Spruce

Sierra Nevada

Death Valley

Oregon pine

Western yellow pine

Douglas fir

Sequoia

Fault Line

Fault Line

1. Atlantic Ocean
2. Chaparral
3. Conifer Forest
4. Desert Areas

of the American deserts, lying between the Cascade range and the Sierra Nevada (which act as a barrier to humid air moving inland from the ocean) on one side and the Rockies on the other. Although this desert has a very uniform appearance, it is in fact inhabited by a considerable variety of plant species.

Most of the animals occur near to natural hollows where water evaporates more slowly. In certain places, such hollows have permitted the formation of vast permanent lakes such as the Great Salt Lake of Utah, and Lake Walker, whose mildly saline waters swarm with animals of various kinds: tiny crustaceans, a host of worm species, numerous frogs and toads, and a variety of fish. The latter attract a multitude of gulls and pelicans—birds which one would not normally expect to see in a desert. To the east, the Basin Region borders on the high plateaux and massifs of Colorado, Utah and Arizona, which are characterized by innumerable canyons. To the south-west, this great desert joins up, via Death Valley, with the Mojave Desert. This zone, characterized by cactuses and yuccas, is inhabited by rattlesnakes, kangaroo-rats, cicadas, tarantulas and scorpions.

Canyons
The photograph *(right)* shows one of the numerous canyons which carve through the North American high plateaux, while the diagrams illustrate the mode of formation of the most famous of them all—the Grand Canyon—in the sedimentary rock of Arizona. The River Colorado, which was responsible for the development of this canyon, initially encountered a zone of almost horizontal rock strata, which had been virtually unaffected by tectonic movements. But on the right bank of the river there was a dislocation which altered the stratigraphic alignment of the rocks. After several tens of thousands of years of eroding action, the River Colorado had burrowed deeply into the rock, leaving on its left bank no more than a few rocky pillars which survived because they were made of more durable rock. The fault to the right of the river, however, had considerably increased the stratigraphic alteration which took place.

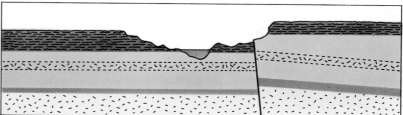

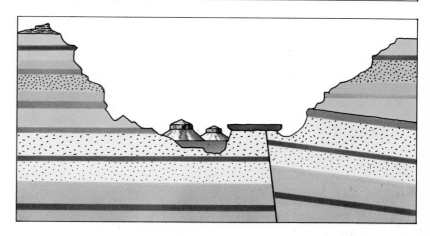

The Basin Region
Most of the animal species found in this region are derived from neighbouring areas. To the north, the climate is not so dry and the more abundant vegetation provides a habitat suitable for rats, shrews and beavers. Raccoons, weasels, otters, mink and foxes can also be found. The southern zone, which is hotter and drier, is inhabited by large numbers of bats, wild rabbits, squirrels, pronghorn antelopes, bighorn sheep and yellow-bellied marmots. Local carnivores include the mountain lion, lynx, coyotes and badgers. Gulls and pelicans are found close to lakes and ponds, while swallows and other insect-eating birds are found wherever insects are available. Raptors prey on both reptiles and rodents, while vultures pick the desert clean of any remains.

Small Predators
Kit-foxes *(above)* and American partridge *(below)* are able to survive thanks to the water they obtain from their prey.

Among the Cactuses of the Gila Desert
Cactuses provide refuge for a variety of animal species. Gila woodpeckers *(above left)* and elf owls make nest hollows in the saguaro cactus. Wrens, goldfinches, linnets and doves nest among the spiny arms of cactuses. Kangaroo-rats make their nests among fragments of cactus and fallen spines. Large predators, on the other hand, are discouraged by this type of vegetation. Wildcats, lynx, coati-mundis, raccoons and skunks will only rarely pursue their prey among the spines.

Tarantulas
Tarantulas (*right:* a black male encountering a brown female), which belong to the trap-door spider group, are large, hairy spiders which can measure up to 25 centimetres (10 inches) across.

Desert Reptiles
In addition to a variety of lizard species which prey upon insects and spiders, the Gila Desert is inhabited by a poisonous reptile, the Gila monster *(above left)*, which has a powerful venom but a relatively primitive mouth structure rendering it less dangerous to man. By contrast, the coral snake *(left)* is a great menace to both animals and man, and the rattlesnake is even more fearsome. *Extreme left:* A desert tortoise.

Swamps and Pine-Groves of the Southern United States

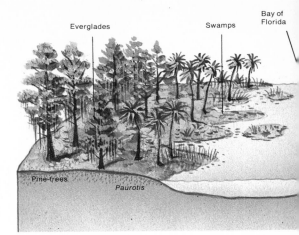

To the west of the Appalachians lies an ancient continental plain which descends through a series of steps to the Gulf of Mexico and the Atlantic Ocean, with a narrow extension into the Florida peninsula. This plain is very largely covered with conifer forest, limited to the west by the Mississippi delta and giving way to swamps along most of the coastline. Behind the coastal sand-dunes lie vast wetland areas inhabited by an incredible variety of animals, most notably bird species. Among other things, there are hordes of tiny rodents belonging to the water-rat subfamily which build spherical nests wedged between three or four reeds, just above the surface of the water. These rodents, which do not swim very well, are extremely acrobatic and are able to move about actively among the reeds without touching the water. They feed on leaves and seeds, supplemented by the occasional slug and a few insects. One of the most interesting zones in this region is the Okefenokee Swamp, which lies between Georgia and Florida and links the Gulf of Mexico directly with the Atlantic Ocean through a series of canals, rivers and small lakes. The dominant trees are deciduous cypresses, which grow either directly in water or in water-soaked soil. The undergrowth is fairly dense and is formed by ferns, climbing plants, laurels and heather. Spanish moss, grey-coloured and bedecked with flowers, hangs in garlands from the branches of trees. The waters of the Okefenokee Swamp are rich in fish, including several pike species and various bowfins, and have a thriving population of alligators, which find a plentiful supply of food there. The adult alligators feed on large fish, while the young ones feed on mussels and slugs. Grass-snakes, a number of other aquatic snake species, turtles and rattlesnakes are also fairly common. Mammals are represented by numerous small rodents, muskrats, grey squirrels, opossums, raccoons, grey foxes, otters, pumas and black bears. Birds are particularly well represented, especially waders such as egrets, herons and sandhill cranes. The Mississippi delta and the Swannee River valley (opening into the Gulf of Mexico at the north-western angle of Florida) are supplied by a number of extremely pure water-sources and both cypresses and oaks are abundant on the banks of these rivers, creating a habitat of great natural beauty. Waders are also very common in these areas: ibis, herons, flamingos and spoonbills. Mammals, such as muskrats, otters, bears and pumas, are present in great numbers. The coastal plain extends from Florida westwards towards the Mississippi, with large expanses of pine-forest. West of the River Apalachicola, which acts as an unexpected faunal boundary, a different fauna is found, characteristic of the Mississippi delta.

The Biological Cycle of the Frog
1. Eggs and freshly-hatched tadpoles, equipped with ventral suckers to attach themselves to aquatic plants.
2. Tadpole with external gills.
3. Tadpole with internal gills.
4. Tadpole with well-developed hind-limbs.
5. Tadpole with all four limbs present; capable of air-breathing, but differing from the adult in possessing a tail.
6. Adult, some four months after hatching from the egg.

Selective Fishing Avoids Competition
The swamps of Florida are inhabited by a variety of heron species which are able to co-exist quite happily. In fact, each species has become specialized for a particular type of feeding behaviour which reduces competition, thus defining separate "ecological niches".

The green heron (1) waits patiently on an emergent root for a fish to pass within reach. The Louisiana heron (2) moves slowly into the water and seeks out its prey. The little egret (3) agitates the water to drive fish out of hiding, whilst the reddish egret (4) first stirs up the water to make the fish swim about and then spreads its wings to mislead its prey into seeking safety in their shade. Finally, the great blue heron (5), which is the largest species of all, has longer legs and can hunt for fish in deeper water where the other herons never venture.

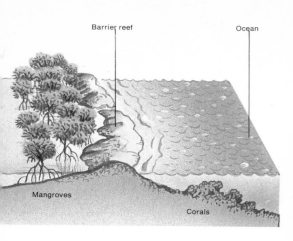

Barrier reef   Ocean

Mangroves

Corals

Florida and the Everglades

The Florida peninsula consists of three fairly distinct areas: the
northern conifer forest zone; the central zone dominated by
prairies; and the southern, largely swampland zone. The
backbone of the peninsula is a ridge of limestone characterized
by sinkholes and by numerous springs. The tip of the peninsula,
where the Everglades are located, is infested by mosquitoes and
is generally foetid. Two seasons can be distinguished: a dry
winter and a wet summer. The whole area lies only just above
sea-level and three successive belts are recognizable. The first
has a network of channels, ponds and small lagoons. The second
is swampy and covered with sedges which dry out in winter. The
third is covered with small hills which are always dry and have a
thriving vegetation of trees and bushes. Alongside the sea is a
red mangrove zone in which a distinctive palm-tree of the genus
Paurotis grows. This is followed by a black mangrove zone which
then gives way to white mangrove which can withstand high
salinity conditions. Finally, some way out to sea, there is a
coral-reef barrier. The ecological cycle of these swampy areas
begins with the mosquitoes which infest the region. They provide
food for small fish of the genus *Gambusia,* which are in turn
preyed upon by perch, garpike, pike and carp. These larger fish
provide food for turtles and alligators. *Above right:* Part of the
Okefenokee Swamp extending into Georgia. *Right:* Swamp-living
cypresses in Florida.

An Aquatic Brigand

The giant diving beetle (shown below in both its adult and larval
stages) is a large-bodied carnivorous species which preys upon
fish, tadpoles and aquatic invertebrates. Both the adult and the
larva are air-breathing. The adult beetle stores air by moving up
to the water surfaces and pushing out the tip of its abdomen to
trap an air bubble beneath the elytra (wing-cases). The larva, on
the other hand, makes use of an abdominal siphon.

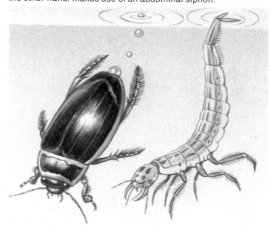

A "Dystrophic" Pond

Although it is richly supplied
with humus (either as a
substrate layer or in
suspension) a "dystrophic"
pond is poor in nutrients
because the breakdown of
organic matter takes place
only slowly and the process is
inadequate to permit
sufficient access to minerals.

The Mississippi Delta

The vast Mississippi Delta region consists of a network of lakes,
rivers, channels and swamps known as "bayous" to the local
inhabitants. Inland from the sea, a succession of vegetation
zones is encountered: littoral, coastal and deltaic. There are, in
addition, three parallel belts of swampland, the outermost with
salt water and the innermost with freshwater conditions. The
dikes and natural dams which delimit areas of flat ground and
water-courses are covered with vegetation which provides
refuge for numerous animal species: otters, herons, egrets and
ospreys. This is also a favoured habitat for a wide variety of frog
and snake species. One of the commonest snakes is the water
moccasin *(above left)*, which feeds primarily on fish and frogs. A
variety of turtle species are also found in the bayous, along with
alligators *(below left)*. But the most characteristic species
inhabiting this region is a rodent of the family Cricetidae—the
muskrat. This rodent is perfectly adapted for aquatic life with
webbed hind-feet, a flattened tail like that of a beaver and an
impermeable skin. It builds exposed nests in summer, but in
winter it digs burrows in the dikes. The bayous are also
inhabited by nutrias—large rodents introduced from the Rio de
la Plata some decades ago, which have adapted well to the local
conditions.

CARIBBEAN SEA

ATLANTIC

OCEAN

LLANOS
Orinoco
GUIANA PLATEAU
Amazon
S E L V A S
CAATINGA
MONTANA
SERTAO
Lake Titicaca
C A M P O S
BRAZILIAN
HIGHLANDS
MATO
GROSSO
ATACAMA
DESERT
GRAN
CHACO
A N D E S
Parana
Parana
Lagõa
dos Patos
Aconcagua
PAMPAS
Rio de la Plata

ATLANTIC

OCEAN

P A T A G O N I A

FALKLAND ISLANDS

TIERRA DEL FUEGO

HUMID TROPICAL

TROPICAL

TEMPERATE
CONTINENTAL

	Hot and humid
	Hot with summer rains
	Semi-arid
	Arid
	Mediterranean
	Continental
	Oceanic
	High montane
	Warm sea-currents
	Cold sea-currents

SOUTH AMERICA

South America was separated from North America by a sea barrier for a long period of geological time. Even today it has conserved distinctive characteristics which are particularly obvious in the immense Amazon basin where an incredible diversity of plants (including more than 4,000 tree species) forms the most luxuriant tropical forest in the world. In this vast habitat area, where gigantic trees are a dominant feature and water is present everywhere, the larger animals have typically become either good swimmers or agile climbers. South of Amazonia and to the east of the Andes lie huge grassland areas which become gradually drier until they grade into the cold subdesert conditions of Patagonia. A transition zone, the Chaco, consisting of deciduous trees, shrubs and

Contrasting Environments in South America
The Peruvian "puna"—a dry, desolate region almost lacking in vegetation—which stretches along the lower slopes of the Cordillera Negra up to an altitude of more than 4,000 metres (13,000 feet), stands in stark contrast to the River Iguaçu, which is fed by vast quantities of water and flows through rich forest vegetation. The River Iguaçu, which runs across the southern high plateaux of Brazil, rushes over a series of impressive waterfalls before flowing into the River Paraña.

swamps, separates the eastern slopes of the Andes and the Amazonian forest from the grassy expanses of the pampas. The Andes run from north to south along the South American continent like an enormous backbone. Between the two parallel chains which make up this mountain range lie vast high plateaux which are on average more than 3,500 metres (11,500 feet) above sea-level. Because of the north-south orientation of the Andes, the vegetation obviously exhibits variation not only with altitude but also with latitude. At tropical latitudes, the Andes constitute a barrier between two climatic extremes: on the west lies the very dry Pacific coastline, with the most arid desert of the world (Atacama), while to the east lies the Amazonian rainforest.

GULF COASTAL PLAIN

FLORIDA

ATLANTIC OCEAN

GULF OF MEXICO

BAHAMAS

CUBA

M E X I C O

GREATER

HAITI

YUCATAN

JAMAICA

ANTILLES

PUERTO RICO

CARIBBEAN SEA

LESSER ANTILLES

PACIFIC

OCEAN

Isthmus of Panama

The Andes

The Andes
The Andes, which run parallel to the Pacific coastline of South America, are of relatively recent origin. The orogenic processes associated with the westward movement of the American tectonic plate still continue today, as is witnessed by volcanic activity and widespread seismic phenomena.

Central America
Central America, a kind of bridge spanning the gap between the two American continents, has a very complex geological structure resulting from volcanic activity. A wide variety of habitat conditions are present. On the Pacific side, the coastline is relatively steep in contour and the environment is comparatively dry, with savannah and thickets of xerophytic plants. The Atlantic coastline, which is exposed to the trade winds, slopes more gently down to the sea and is covered with a rich vegetation including mahogany, bracken, epiphytes and lianes.

The West Indies
The archipelago which Columbus christened the "West Indies" actually consists of the peaks of a mountain-chain which has been largely submerged in very deep water. The very fertile soil, often of volcanic origin, the uniformly high temperatures and heavy rainfall combine to encourage the growth of luxuriant, extremely varied vegetation. On the various Caribbean islands, savannah is found up to altitudes of 600 metres (2,000 feet), giving way first to dense forests of laurels, conifers and palm-trees, which extend up to an altitude of 1,000 metres (3,300 feet), and then to a zone of perennial plants and heaths which constitute the montane prairies.

The Andes have an extremely complex structure in Colombia; they are formed by two large chains which divide into three distinct branches south of the Colombian massif. The two long, narrow valleys separating these branches are in fact two rifts in the Earth's crust, analogous to the African Rift Valley. The north-eastern slopes, which lead down to the Caribbean Sea, are dry, whereas the Pacific slopes of Colombia and Ecuador have a high rainfall throughout the year. It is this which has allowed the development of dense rainforest. The rainforest extends up to an altitude of 1,800 metres (6,000 feet) before giving way to subtropical forest and then, higher up, to temperate forest. From 2,500 to 3,500 metres (8,000 to 11,500 feet) the vegetation becomes progressively sparser until the trees are entirely replaced by "paramos", or green prairies. Further south, these prairies grade into the steppes which characterize the high plateau of Peru. South of this dry zone, the Andes bear vegetation similar to the Mediterranean maquis. This is followed by forests of *Araucaria* pines (monkey-puzzle trees) and, in the south, by beech and cedar forests.

Anoles
Anoles, or American "chameleons", are typical inhabitants of tropical areas of the New World. They are reptiles belonging to the family Iguanidae and the males are characterized by a brightly-coloured throat-sac which can be extended by means of a complex mechanism. Anoles are agile climbers and their digits bear tiny transverse ridges with an adhesive action.

A Naturalist's Paradise
The Amazonian rainforest, which persists up to an altitude of 1,000 metres (3,300 feet), the tropical deciduous forests and the leafy maquis of the Pacific coast together provide habitats for a tremendous variety of animal species, notably for birds (including many migratory forms which stop over in this region in the course of their travels). The coati-mundi *(above)*, a typical inhabitant of Central America, is a carnivore belonging to the family Procyonidae. *Below:* A large tarantula *(Sericopelma communis)* which has just captured a humming-bird.

St. Lucia Parrot

The Caribbean Parrots
The various parrot species which live on the Caribbean islands are almost certainly derived from Central America originally. They prefer to live in the wettest areas, in forests with an abundant food supply. *Right:* A typical Peruvian village.

Hispaniolan Parrot

The Andean Flamingos
Several flamingo species are native to the Andean region. These birds depend upon water, where they seek out their food consisting (according to the species) of plant fragments or tiny molluscs and crustaceans. The flamingo's beak is equipped with fine plates (lamellae) which enable it to filter out small food items from the water, with the tongue acting as a piston to suck up water or eject it from the beak (see below).

The Andean Condor
Although now very rare, the powerful Andean condor (above right) actually has a vast area of distribution. These birds not only occur in the Andes but also fly above the beaches of Peru in search of dead fish and sealions.

Running Birds
Rheas (right) are typical birds of the South American grasslands and the high plateaux of the Andes. In both behaviour and appearance they closely resemble the ostriches of Africa, though they are only distantly related to them in evolutionary terms. Rheas are unable to fly, but they are able to run very quickly across the grasslands where they forage for grasses and herbaceous plants.

The Animals of the Andean High Plateaux
Because of the considerable altitude of the Andean high plateaux (exceeding 3,500 metres, or 11,500 feet), the vertebrates found there have better-developed hearts and lungs than their plains-living relatives. Their blood is richer in haemoglobin in order to offset the rarefication of the atmosphere and to supply adequate quantities of oxygen, which is indispensable for body metabolism. Below: A group of llamas, relatives of the camel which are typical inhabitants of the Andes. Bottom: A Bolivian landscape.

The Peruvian Puna
Above the level at which thickets of xerophytic plants are found, there is a steppe zone with scattered bushes. This is followed by a high montane prairie zone at altitudes in excess of 3,500 metres (11,500 feet). This semi-desert region is characterized by perennial plants adapted for extremely dry conditions, marked exposure to sunlight and a rarefied atmosphere. Three types of vegetation are predominant: cushion-like plants, rosette-forming plants and dwarf bushes. Below: Three examples of plant species found at high altitudes: Lobivia inguiensis, Lotamphocereus otuscensis and Vinteria auresbeina.

The Andean Lakes
Numerous lakes are found on the high plateaux of Peru and Bolivia, some of them reaching considerable dimensions. These lakes exert a thermo-regulatory influence, since their waters have a constant temperature year-round. Lake Titicaca (above), which lies at an altitude of 3,800 metres (12,500 feet) is 203 kilometres (127 miles) long and 65 kilometres (41 miles) wide. The water temperature is always between 10 °C and 12 °C (50 °F and 54 °F) and its rush-lined shores provide refuge for a variety of animals, particularly birds. There are grebes, giant coots, cormorants, gulls, flamingos and an amazing diversity of duck species. Below: An entirely different kind of landscape—the Beagle Channel of Tierra del Fuego, at the extreme southern tip of South America.

Horned coot

Short-winged grebe of Lake Titicaca

The Amazonian Rainforest

Because of the abundant availability of water, the Amazonian vegetation has a luxuriance which is unmatched anywhere else in the world. The humidity level is so high that some animals belonging to groups which are habitually water-living, such as crustaceans, leeches and frogs, lead a terrestrial existence in the Amazon basin. The high ambient temperature also seems to favour evolutionary diversification, particularly among the invertebrates which are by far the most numerous in terms of species. Another special feature of the forest is that it contains "giant" beetles, butterflies and spiders. The Hercules beetle *(Dinastes hercules)*, characterized by the long "horn" on its carapace, is 20 to 22 centimetres (8 to 9 inches) in length, whilst some cockroaches are more than 10 centimetres (4 inches) in length. Tarantulas—large hairy spiders—can measure up to 25 centimetres (10 inches) across with their legs spread out. Ants play a major role in the forest economy by transforming organic material. Differential distribution of light in the forest has an influence on the coloration of animal species living at different levels. Almost all the animals in Amazonia are either arboreal or aquatic. The undergrowth is inhabited by white-tailed deer, which are excellent swimmers, and various peccary species which feed on fruits, corms and roots. There is also a giant anteater which traps ants and termites with its long, sticky tongue. Food is obviously more abundant high up in the trees where the sun can penetrate; accordingly the greatest number of animals is to be found there. The larger species all have slim, agile bodies, bright colours and strong claws or grasping extremities to permit a good grip on the branches. Monkeys, tamanduas, coendus (porcupines with prehensile tails,) kinkajous and sloths are particularly well-equipped and all except the sloths have prehensile tails.

The Vegetation
Characteristic plants of the rainforest include the rubber-tree *(Hevea)*, the castilloa, various palm-tree species and the cacao-tree. The rainforest forms a vast block extending along the lower Amazon basin.

Bees
Bees are particularly common in Amazonia, where they live in typical matriarchal societies in which only the queen is able to produce eggs. Mating takes place on the wing and only the fittest male will be able to leave offspring by fertilizing the queen. Other females (workers) construct the cells of the hive and subsequently produce honey and the royal jelly. Worker females are, in fact, individuals which receive only a limited food-supply and sacrifice their own chances of breeding for the good of the hive. *Left:* A section through a typical bee-hive.

Pollen cells

Honey cells

Cells for rearing of larvae

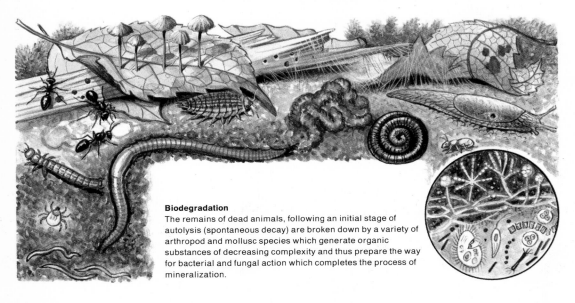

Biodegradation
The remains of dead animals, following an initial stage of autolysis (spontaneous decay) are broken down by a variety of arthropod and mollusc species which generate organic substances of decreasing complexity and thus prepare the way for bacterial and fungal action which completes the process of mineralization.

South American Hunting Wasps
Certain South American wasp species are able to capture large tarantulas, which are paralyzed with venom from the sting and subsequently provide food for the wasp's developing larva.

Canopy-Living Animals
Arboreal mammal species predominate in the Amazonian rainforest. In addition to a large number of New World monkey species, such as the squirrel-monkey *(above)*, the uakari *(right)* and the woolly monkey *(below right)*, there are tree-living tamanduas, coendus (prehensile-tailed porcupines), and a number of marsupials such as the thick-tailed opossum *(centre of page)*. Large numbers of birds with multi-coloured plumage make this habitat a paradise for the naturalist. Quetzals *(below)*, toucans *(below right)* and parrots feed on fruits, while humming-birds suck nectar from flowers. The Amazonian rainforest also provides a home for numerous insectivores which feed upon the abundant insect fauna.

The River Amazon
The River Amazon, with a basin covering more than seven million square kilometres (three million square miles), constitutes the greatest river drainage system of the world. The primary sources of the Amazon are in Peru and the river flows eastwards, almost parallel to the equator, with only a very slight downward slope. Surrounded by an extremely dense expanse of rainforest, the river flows across the low Amazonian plain, which is constituted by alluvial sediments deposited on the extremely ancient rocks of the Brazilian shield. The affluents of the River Amazon, some of which—such as the Rio Madeira—are longer than the great European rivers, are symmetrically distributed on either side and all of them carry enormous quantities of detritus into the Amazon.

A Powerful Poison
The skin of tree-frogs of the genus *Dendrobates (right)* secretes a poison comparable to curare, which is used by Amazonian Indians on the tips of their poison arrows.

Mammals with armour
Armadillos *(below)* are the only mammals equipped with a coat of armour. They are burrowing mammals which feed upon insects and snakes.

The Fauna of the Amazon Basin
The freshwater habitat of the Amazon region is inhabited by an extremely rich fauna dominated by fish, believed to number some 2,500 species. The largest of these is the arapaima, which can reach five metres (16 feet) in length and weigh as much as 100 kilograms (220 pounds). The widely feared piranha is also found, along with electric eels (capable of producing a discharge of 350 volts) and rays of the genus *Potamotrygon*, which can inflict painful wounds with their whip-like tail equipped with small spines linked to poison glands. There is even an Amazon dolphin *(Inia)* which feeds on fish. Reptiles are represented by numerous turtle and cayman species, the most dangerous being the black cayman, which can reach a length of up to six metres (20 feet).

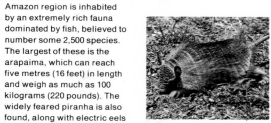

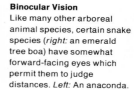

Binocular Vision
Like many other arboreal animal species, certain snake species *(right:* an emerald tree boa) have somewhat forward-facing eyes which permit them to judge distances. *Left:* An anaconda.

Undergrowth Animals
Tapirs *(left)*, otters and large rodents such as agutis, nutrias and pacas are all excellent swimmers although they are essentially terrestrial mammals. *Right:* A jaguar, the biggest carnivore in Amazonia, weighing up to 250 kilograms (550 pounds) and measuring up to 2,5 metres (8 feet) in length.

77

From Desert to Forest

The vast high plateaux which form a long corridor between the two main chains of the Andes constitute a unique habitat at an altitude of 3,500 to 4,300 metres (11,500 to 14,000 feet). To the west lie the deserts of the Pacific coastal region; to the east lies the immense rainforest belt. Despite their tropical latitude, the Andean high plateaux have a low average temperature of only 2 °C (36 °F). During the summer, rain falls almost every day, but the climate is still quite dry because of the strong winds which rapidly drive off any water. The vegetation is composed of dwarf plants, perennial grasses and rosaceous species. The high plateaux of Venezuela, Colombia and Ecuador are carpeted with "paramos"—mountain pasturelands with a cool, humid climate, inhabited by woolly tapirs, spectacled bears, hummingbirds, ovenbirds, thrushes, finches and wrens. The high plateaux of Peru are predominantly covered by the grassy steppe known as "puna", which is inhabited by various hardy plant species. Near to the water-courses live guineapigs, llamas, vicuña, numerous rodents (such as viscachas) and Andean foxes in addition to a variety of birds including urubus (vultures) and condors. To the south of the Amazonian rainforest block lies a vast savannah zone including wooded areas, and marshland gradually gives way to the pampas. In Patagonia, the prairies become progressively more arid and on the Atlantic side there are subdesert conditions; this is the home of guanacos, vicuñas, various deer species and numerous ducks.

A Mosaic of Biological Zones
The Andes, with their three main mountain-chains, bear large expanses of rainforest which gradually disappears with increasing altitude to give way to dry prairies (Peru) or to "paramos" covered with a flora consisting of composite plants, umbelliferous species and gentians.

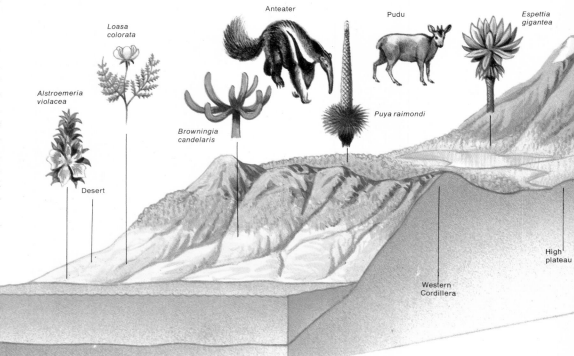

Alstroemeria violacea

Loasa colorata

Browningia candelaris

Anteater

Puya raimondi

Pudu

Espettia gigantea

Desert

Pacific Ocean

Western Cordillera

High plateau

Arid Coastal Deserts, But Rich Marine Life
Along the Pacific coast stretches a narrow band of rocky or sandy desert lands which extend from 3 ° to 30 ° latitude over a distance of about 3,500 kilometres (2,200 miles). Only here and there is the desert landscape interrupted by patches of vegetation, usually situated close to torrents descending from the Andes. The climate is characterized by relatively cool conditions (about 20 °C, or 68 °F), with restricted rainfall but relatively high ambient humidity. Despite the relatively low latitudes, the sea is very cold because of the Humboldt current which transports the icy waters from the south towards the equator. The air which comes into contact with the cold sea-water cools more rapidly than that in the upper layers of the atmosphere, with the result that a belt of fog is generated at a height of 300 to 1,000 metres (1,000 to 3,300 feet), blocking out much of the sunlight. Only relatively few animal species are to be found here. Lizards and geckos prey upon the few insects and scorpions, whilst rodents seek out seeds and grasses. Further south, along the northern coast of Chile, lies the most arid desert of the world, Atacama (left). Here, there are only a very few patches of vegetation, limited to the lowest areas where underground water is present and composed primarily of plant species of the family Mimosaceae. This extreme aridity stands in stark contrast to the richness of life in the sea, for the cold waters of the Humboldt current carry an abundant supply of plankton which provides food for a multitude of plankton-feeding fish, such as anchovies, which in turn provide food for larger fish. The food-chain is completed by the seals and whales, along with the vast numbers of birds which nest along the coast: gulls, pelicans, petrels and scissor-bills.

The Rainforests of Guyana and Brazil

A large part of Guyana and eastern Brazil is covered with a dense expanse of forest which extends up to an altitude of 3,000 metres (10,000 feet). According to their particular dietary habits, birds are found at different levels in this forest. Ibis, egrets, herons and ducks remain close to water-courses, while the cock-of-the-rock, the great razor-billed curassow and the red tinamou, which feed on fruits and insects, nest on the forest floor. Most species, however, live up among the foliage. Woodpeckers, barbets, trogons and parrots all nest in tree-hollows. A distinguishing feature of rainforest bird species is their sedentary habit, with each individual occupying a relatively small home range, reflecting the abundance of food and ready availability of refuges. Some bats feed on rodents or fish, while others suck the blood of large mammals. Yet others feed on fruit or on the nectar produced by flowers. The predators of the rainforest include caymans and snakes, such as the very poisonous rattlesnake and strangling species including the anaconda and boa constrictors. Birds of prey are well represented; they include the huge harpy eagle, a savage predator of monkeys and sloths, and numerous vultures. Mammalian predators include jaguars, ocelots, tiger cats, jaguarundis and tayras.

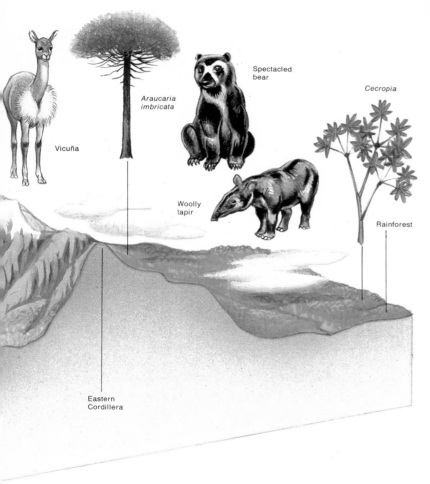

Vicuña

Araucaria imbricata

Spectacled bear

Cecropia

Woolly tapir

Rainforest

Eastern Cordillera

The Mato Grosso

Inland from the coast, annual rainfall tends to diminish. For example, in the centre of Brazil there are large dry central plateaux where the dominant vegetation consists of spiny trees and other xerophytic plants, sometimes forming an impenetrable barrier. The animal life of such areas is somewhat limited and includes armadillos, a wide variety of rodents, a few monkeys (such as small marmosets), carnivores including hog-nosed skunks, and numerous birds such as the golden parrot, a number of woodpecker species and a variety of passerines. The Mato Grosso, which is characterized by a marked alternation of dry and wet seasons, is a vast expanse of savannah-type grassland with patches of trees growing here and there. Although the region is therefore rather monotonous in appearance, a variety of habitat conditions can be found, ranging from thick forest through swampland to desert zones. In addition to enormous flocks of parrots and doves, there are also groups of rheas to be found. The local mammals include deer species, tapirs, peccaries and jaguars. The vegetation includes relatives of the water-lily (family Nymphaceae) such as *Victoria regia (above)*.

From One Coast to the Other

The schema *(left)* represents the profile of South America in a section taken from one side to the other. It is interesting to note that large lakes are present on the Andean high plateaux. The high altitude of these lakes and the resulting low temperatures explain the presence of specially adapted amphibians which have developed a cornified protective layer over their skins. In some cases, as in Lake Titicaca, these animals have completely returned to aquatic life.

The Pampas

The pampas are vast expanses of treeless steppe covering a total area of almost 780,000 square kilometres (300,000 square miles) extending from the Atlantic to the foothills of the Andes *(above)*. They are constituted by an accumulation of alluvial red earths carried down by river torrents from the Andes and they have been considerably modified by cattle-ranching. Birds are the most prominent members of the fauna, including many migratory species.

The High Plateaux of Peru

The high plateaux in Peru are dry and covered with steppe. Characteristic animal inhabitants include llamas, vicuñas and a large number of rodent species.

The Pantanal, the Chaco and the Subantarctic Region

Between the high plateaux of Brazil and Bolivia lies the Paraguayan valley known as the "Pantanal", which has a network of marshes which attract a multitude of birds: rails, spoonbills, ibis, jabirus, cormorants, ducks and others. Another, less well-known, marshland region is the Chaco, which is located on the Bolivian slopes of the Andes. This region, which is difficult to reach, is covered with a mosaic of ponds and dry areas. The vegetation varies accordingly, ranging from forests on the mountain slopes to savannah on the plains. Further south, the steppe becomes increasingly arid and eventually grades into the subantarctic zone, which is shown *(right)* with a colony of Magellanic penguins.

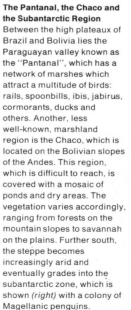

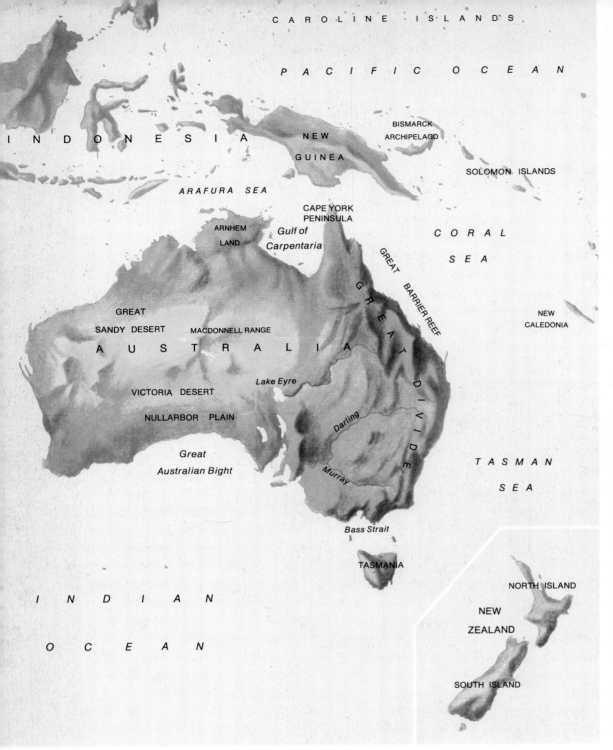

CAROLINE ISLANDS

PACIFIC OCEAN

INDONESIA

NEW
GUINEA

BISMARCK
ARCHIPELAGO

SOLOMON ISLANDS

ARAFURA SEA

CAPE YORK
PENINSULA

CORAL

SEA

ARNHEM
LAND

Gulf of
Carpentaria

GREAT BARRIER REEF

NEW
CALEDONIA

GREAT

SANDY DESERT   MACDONNELL RANGE

AUSTRALIA

GREAT DIVIDE

VICTORIA DESERT   Lake Eyre

NULLARBOR PLAIN

Darling

TASMAN

SEA

Great
Australian Bight

Murray

Bass Strait

TASMANIA

NORTH ISLAND

NEW

ZEALAND

INDIAN

OCEAN

SOUTH ISLAND

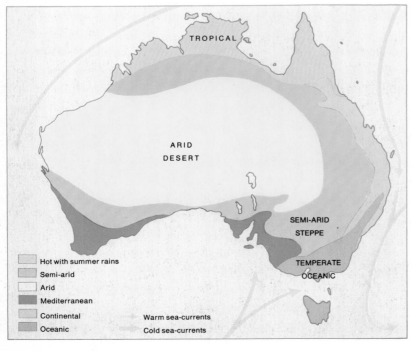

TROPICAL

ARID
DESERT

SEMI-ARID
STEPPE

TEMPERATE
OCEANIC

	Hot with summer rains
	Semi-arid
	Arid
	Mediterranean
	Continental	Warm sea-currents
	Oceanic	Cold sea-currents

AUSTRALASIA

Australia is the flattest continental land-mass. This enormous island only exceeds an altitude of 700 metres (2,000 feet) over less than five per cent of its area. Its highest peak, Mount Kosciusko, in the southern part of the mountain-chain known as the Great Dividing Range, is barely 2,500 metres (8,250 feet) high. Three major zones can be distinguished: the forested zone, the arid interior and the northern tropical region. The most striking feature overall is the low rainfall: an average annual figure of only 25 centimetres (10 inches)! This is largely due to the presence of the Great Dividing Range, which blocks off the humid trade-winds from the Pacific, and to the marine currents from the South Pole which exert a cooling and drying effect on winds arriving from the west. As a result, the entire central region of Australia is characterized by sandy deserts with patches of mulga scrub (mulga is a kind of acacia) and spiny plants here and there. The centrally located Mac-

A Landscape in the Australian Alps of South-East Australia
Like the other mountain-chains in eastern Australia (Blue Mountains; Great Dividing Range), the Australian Alps represent the much-diminished vestiges of extremely ancient Palaeozoic folding events. The mountain-chains of New Zealand are far more recent in geological terms and they incorporate numerous lakes of glacial origin. *Right:* Pukaki Lake.

Donnell Ranges have an altitude of between 300 and 500 metres (10,000 to 16,500 feet). These ranges have some very interesting peculiarities, since the survival of certain plant species, such as palm-trees and sago-palms, demonstrates that this region once had a plentiful supply of water. In the ponds, there are crayfish, aquatic insects and fish which are entirely lacking in the surrounding desert. To the north-east lies a huge arc of dry prairie-land which constitutes a transition zone between the desert region and the coastal forest. To the east lies a zone of wooded savannah which precedes the true forest and extends down along the entire southern region. But there is relatively little forest in Australia; it is confined to a comparatively narrow belt which is best developed in the east, between the mountains and the coastline. Forest is also present in the south-western part of Australia and in Queensland, where it becomes typical rainforest.

Deserts and Prairies

A third of Australia is covered with hot deserts. Most are vast expanses of red sand which in the east form fixed dunes arranged in parallel lines, though in the south stone deserts predominate. In the central region lie numerous dried-out lake beds covered with a thin layer of crystallized gypsum which gives off a reflection and thus produces the illusion that the lakes are still full of water. The Australian deserts contain many bizarre animals, including marsupials, both insectivores and a range of herbivorous species. The marsupial mole, very similar in appearance to the European mole, inhabits the driest regions, where it constructs complex burrow systems and seeks out insect prey. For all the desert animal species, the most serious problem is that of obtaining water. Frogs have developed a special mechanism to cope with the arid conditions: they store water in a specially adapted body-chamber, which is blown up like a ball, and they secrete a capsule of mucus around themselves before burrowing into the mud at the bottom of water-pools. In this way, such frogs can survive for as long as two years imprisoned in the dry mud of a dessicated swamp. When the latter eventually fills up with water again, the frogs make their way to the surface to breed. The metamorphosis of the tadpoles must take place during the few weeks in which water is available. Crayfish, crabs and gobies show similar behaviour which allows them to survive long periods of dessication. In north-eastern Australia, where rainfall is higher, the desert gives way to grassland, which is particularly extensive in Queensland and is limited to the east by the Great Dividing Range. To the south, a large sector of this grassland zone forms the basin of the Murray and Darling Rivers. Majestic eucalyptus trees are found here and the rivers are inhabited by hake, blackfish, golden perch, gobies and gurnets.

The Australian Vegetation
Because of the long isolation of the Australian continent, the flora contains a large proportion of endemic species which are found nowhere else in the world. Conversely, plants which are common elsewhere are relatively rare in Australia. The endemic plant species of Australia include (in addition to the well-known eucalyptus): beefwoods, sago-palms, screw-pines and papaw trees. The interior, which has very low rainfall, is characterized by three main types of vegetation: 1. Savannah with very few trees; 2. Scrubland, typified by acacia bushes; 3. Steppe, with various grass species and a few bushy acacias, which gives way to desert in places where the conditions are particularly dry. The photograph *(above)* illustrates the dry, monotonous character of the Australian savannah.

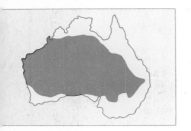

The Australian Deserts
A very large part of Australia is characterized by dry climatic conditions (less than 30 centimetres, or 12 inches, annual rainfall and marked variation in temperature). This explains the large area covered by deserts, especially in the central region where the sea exerts virtually no influence at all. The deserts vary in appearance. Some are stony deserts with isolated granitic or quartzitic outcrops, which are buffeted by strong winds, such as the well-known Ayer's Rock in the Northern Territory. There are also sandy deserts with widespread dunes and occasional brackish ponds.

Terrestrial Birds
The emus *(above)* are large birds which are unable to fly but can run very fast, like the African ostriches. They live in small groups of about a dozen individuals and they feed on berries, grasses and insects.

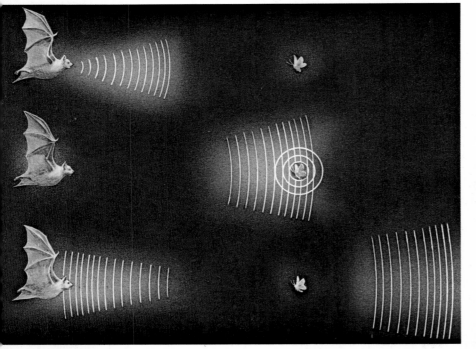

Natural Echolocating Mechanisms
Echolocation, which is used by microchiropteran bats to orient themselves, to avoid obstacles and to locate their prey, is a natural sonar system. When the conical beam of ultrasound emitted by the bat (shown in blue in the diagram) encounters an object, the sound waves are reflected (shown in pink) and the bat's ears pick up the echo.

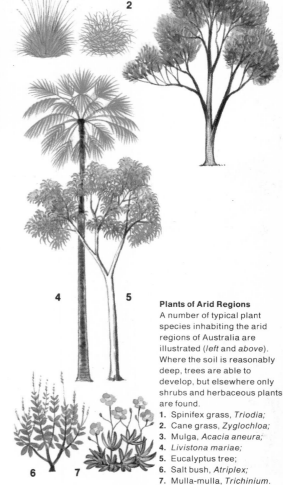

Plants of Arid Regions
A number of typical plant species inhabiting the arid regions of Australia are illustrated *(left* and *above)*. Where the soil is reasonably deep, trees are able to develop, but elsewhere only shrubs and herbaceous plants are found.
1. Spinifex grass, *Triodia*;
2. Cane grass, *Zyglochloa*;
3. Mulga, *Acacia aneura*;
4. *Livistona mariae*;
5. Eucalyptus tree;
6. Salt bush, *Atriplex*;
7. Mulla-mulla, *Trichinium*.

The Scorpion Dance
Scorpions are nocturnal in habits and spend the day in deep burrows excavated in the ground. Mating involves a peculiar ritualized "dance" *(above)*. The male deposits a spermatophore on the ground and the female retrieves it with a special sequence of movements, as the partners circle around with their pincers locked together.

Desert Life
Despite the hostility of the environment, the Australian deserts are inhabited by a rich variety of animals, most of which lead a nocturnal existence to escape the high daytime temperatures. Mammals are represented by dingos, bandicoots, rabbits, rat kangaroos, crest-tailed marsupial mice and wallabies. In areas with somewhat richer plant growth, red kangaroos are also found. Birds are well represented: falcons, long-tailed eagles, kestrels and kites are able to find adequate food because of the abundance of reptiles and small marsupials in the desert; finches, parrots and pigeons are able to find enough seeds and fruits to survive.

Frilled lizard

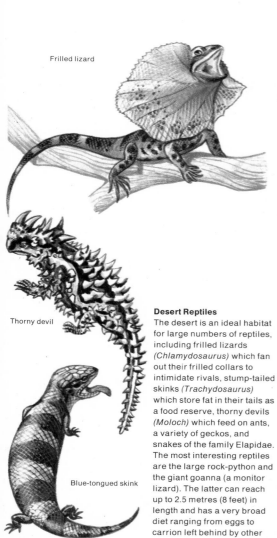

Thorny devil

Blue-tongued skink

Desert Reptiles
The desert is an ideal habitat for large numbers of reptiles, including frilled lizards *(Chlamydosaurus)* which fan out their frilled collars to intimidate rivals, stump-tailed skinks *(Trachydosaurus)* which store fat in their tails as a food reserve, thorny devils *(Moloch)* which feed on ants, a variety of geckos, and snakes of the family Elapidae. The most interesting reptiles are the large rock-python and the giant goanna (a monitor lizard). The latter can reach up to 2.5 metres (8 feet) in length and has a very broad diet ranging from eggs to carrion left behind by other predators

Birds of the Plains
Birds are the most prominent animals in western Australia: kites, emus, sulphur-crested cockatoos *(left)* and galahs *(above)*, parakeets and lorikeets. In the temporary marshes formed by the flood-waters of the rivers during the rainy season, ibis, pelicans, cormorants, egrets, bitterns, black swans, coots and ducks can also be found.

The Red Kangaroo
Kangaroos are social mammals. They live in groups dominated by an adult male. During the day, they spend their time grazing; the hottest hours of the day are spent resting in the shade of bushes. They can attain a running speed of 50 kilometres per hour (30 m.p.h.) and their leap can be as long as eight metres (26 feet).

A Colony of Social Insects
Termites are insects which have developed a complex pattern of social organization based on division of labour, involving the differentiation of individuals into castes: fertile individuals (winged queens and kings), soldiers and workers. Some species construct underground nests while others build mounds which may be up to eight metres (26 feet) high. Such mounds are extremely strongly built. The inside humidity and temperature are maintained constant to provide a stable microclimate indispensable for the well-being of the colony. Termites feed on decaying vegetation which they are able to digest thanks to symbiotic bacteria which live in the intestine and are able to break down cellulose. Large-bodied termites of the genus *Macrotermes* in Africa and tropical regions of Asia cultivate fungi. In order to do this, they prepare culture gardens in their mounds with plant fragments carried in by the workers. The fungi which grow on the fragments transform the lignin of wood into accessible nutrients. There are more than 150 termite species in Australia, some of which are very primitive.

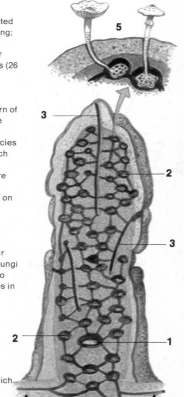

Section through a Termite Mound
1. Royal chamber (copularium); **2.** Fungus garden chamber; **3.** Aeration chimney; **4.** Tunnels leading to the dead wood which provides the food; **5.** Detailed close-up of a fungus chamber, showing the development of fruiting bodies of the fungus.

New Guinea
Heavy rainfall distributed fairly evenly throughout the year and the high temperatures typical of equatorial regions combine to support dense rainforest throughout New Guinea. The only variation in climatic conditions on this island is produced by changes in altitude. Thus, montane prairies and conifer forests can be found in the centre of the island, where there is a series of longitudinally oriented mountain-chains, of Tertiary origin, exceeding 5,000 metres (16,500 feet) in height.

Blue bird of paradise

Lesser superb bird of paradise

Little king bird of paradise

Lesser bird of paradise

Forests and Woodland

In the eastern part of Australia, the forest forms a narrow belt between the mountains and the coastline. There is also a small area of forest in the neighbourhood of Adelaide. The southern coast is extremely varied in appearance; some areas are covered with dense vegetation, while in drier areas bracken, mountain ash and laurels are found. In the forests of the State of Victoria, where eucalyptus trees predominate, there are a number of peculiar animal inhabitants such as the large black cockatoos, arboreal phalangers and the big grey kangaroo. The most unusual species of all is the giant earthworm, which may be as much as three metres (10 feet) in length and produces deep gurgling sounds when it burrows into the light soil in the undergrowth. Temperate forest is found in the south-western region. Here, the vegetation is lush and largely constituted of plant species which are not found elsewhere, since the desert provides a considerable biological barrier. By contrast, the fauna closely resembles that of the south-eastern temperate forest, which also covers a large part of Tasmania. It is generally agreed that the latter island is effectively a part of Australia which was cut off by a rise in sea-level. The main plant species of this forest are eucalyptus and Antarctic beech, while the climate varies from temperate to cold and rainfall is quite heavy. Local mammals include native "cats", opossums, kangaroos, Tasmanian devils and spiny anteaters. Some birds, such as petrels, albatrosses and penguins, visit the Tasmanian coast while migrating, whereas others are permanent residents of the island: honey-eaters, parrots, swallows and pigeons. Large populations of black swans are found in the lagoons. In the Queensland rainforest, the largest in Australia, the fauna and the flora exhibit great similarities to those of New Guinea. Thick, humid jungles line the slopes of the mountains, while screw-pines and palm-trees fringe the beaches.

The New Guinea Fauna
New Guinea, which is characterized by a central mountainous backbone exceeding 5,000 metres (16,500 feet) in altitude, has a large diversity of climatic conditions and habitats, such as the tropical swamp (right). Above the lowland tropical forest zone lies montane forest which eventually gives way to vast prairies. This island is the home of the birds of paradise (above), of which there are some forty species ranging in size from that of a sparrow to that of a pigeon. There are also cassowaries (left), honey-eaters, bee-eaters, parrots and an unusual cockoo species—the coucal—which actually builds its own nest and incubates its own eggs. In the coastal mangrove zone live kingfishers, falcons, sandpipers and egrets, all of which feed on crabs. Reptiles are also abundant: there are monitor lizards, crocodiles, freshwater turtles and various snakes such as the death adder and the well-known taipan. As in Australia, placental mammals are represented only by rodents and bats, whereas marsupials are very common.

84

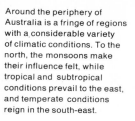

Around the periphery of Australia is a fringe of regions with a considerable variety of climatic conditions. To the north, the monsoons make their influence felt, while tropical and subtropical conditions prevail to the east, and temperate conditions reign in the south-east.

Solar Incubation

Gravid female sea-turtles approach the beaches during the night and dig a hole in which they lay up to a hundred eggs, which are subsequently covered with a layer of sand. After egg-laying, the females return to the sea and incubation is carried out by the sun's heat. Eight weeks later, the tiny turtles hatch and head towards the sea, guided by an infallible instinct. During their brief journey to the sea, they are preyed upon by Dominican gulls, sand goannas and even ghost-crabs. All these predators have such voracious appetites that only a small proportion of hatched turtles actually reach the sea.

New Zealand

New Zealand, lying some 2,000 kilometres (1,250 miles) to the south-east of Australia, is composed of two large islands separated by the Cook Strait. North Island has a subtropical climate and is characterized by volcanoes, hot springs, geysers and high plateaux covered with ferns and shrubs. South Island is characterized by deep, picturesque fjords, high mountains with glaciers and cold-water lakes, and luxuriant forests of Antarctic beech. New Zealand has a rich fauna, particularly noteworthy for its unusual bird species. Many bird species, including albatrosses, gulls and oystercatchers, are Antarctic migrants. Endemic species include a number of flightless forms such as the kiwi *(below)*, the weka, the purple gallinule and the kakapo (a parrot species). This is partly explained by the fact that there are no mammalian predators in New Zealand (apart from bats), with the result that birds have occupied certain terrestrial ecological niches. On the high plateaux of South Island lives the kea, the only meat-eating parrot in the world.

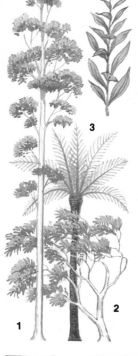

Eucalyptus Trees

In addition to tree-ferns, which are also encountered in other areas with a moist climate, Australian regions which receive heavy, uniformly distributed rainfall have characteristic forest growth dominated by eucalyptus trees. These forests, which may be dense or relatively open, are known to contain about six hundred eucalyptus species, many of which are rich in essential oils employed either in medicine or in the manufacture of cosmetics. Because of their very rapid growth, eucalyptus trees have also played an important part in reafforestation schemes in various parts of the world.
1. Giant eucalyptus;
2. Snow-gum;
3. Tree-fern;
4. Waratah.

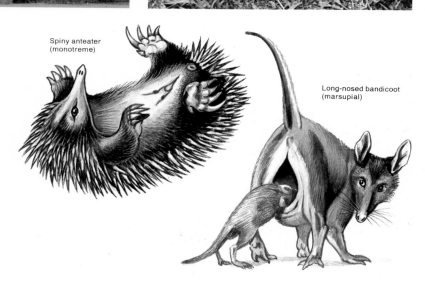

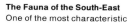

The Mammals

The world's mammals belong to three quite distinct evolutionary groups: the monotremes, the marsupials and the placentals. The main feature shared by all three groups is suckling of the offspring by the mother, but they differ in the way in which the offspring actually develop. The monotremes lay eggs in which the embryo has already begun to develop. Marsupials give birth to very poorly developed offspring which typically complete their development in the mother's pouch. In placentals, the offspring is nourished by a placenta prior to birth and is born at a relatively advanced stage of development.

Spiny anteater (monotreme)

Long-nosed bandicoot (marsupial)

The Fauna of the South-East

One of the most characteristic species of this region is the koala *(above)*, which looks rather like a young bear but is actually a tree-living marsupial living in close association with the eucalyptus trees whose leaves it eats. In these same temperate forests there are also opossums, grey kangaroos and native "cats". Birds are very numerous, particularly the honey-eaters. These small to medium-sized birds have a flattened tongue with which they extract pollen and nectar from flowers, incidentally assisting the latter in cross-fertilization. Insectivorous birds include tree-creepers, warblers and swallow shrikes. Forest king-fishers of the genus *Halcyon* catch fish by diving into rivers or ponds, while the closely related kookaburra stays in the foliage to hunt for insects, snakes and lizards. The very attractive lyre-birds *(above)* feed on crustaceans and insects.

85

The Arctic

The Arctic
Because of the poor availability of mineral salts and the presence of ice over the sea for a large part of the year, phytoplankton (consisting of microscopic algae) develops only slowly in the Arctic. In summer, however, with the influx of warm water from the Atlantic bearing large quantities of mineral nutrients and with almost uninterrupted sunlight, there is a flush in these small plants which provide the basis for the entire ecological cycle in this habitat. The zooplankton, which feeds upon the phytoplankton, in turn provides food for shrimps, sponges, starfish and other vertebrates. These are preyed upon by fish, pinnipeds (seals and sea-lions) and toothed whales. Baleen whales, on the other hand, feed directly on the zooplankton by filtering the sea-water. The food-pyramid of the Arctic is completed by a variety of predators, the most ferocious being the killer whale in the water and the polar bear on the pack-ice.

Arctic Carnivores
Above: A polar bear; the largest carnivore of the Arctic region. *Left:* An Arctic fox. Although the coat of this fox is white in winter, it becomes grey-brown in summer following the moult.

In both males and females the canine teeth of the walrus serve a variety of functions: as weapons for fighting; as tools for probing the sea-bed in search of food (molluscs and crustaceans); and as picks to assist movement across the ice.

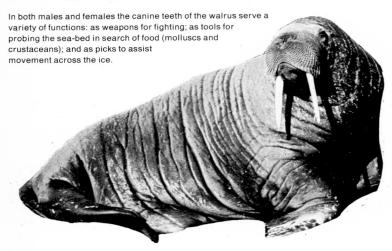

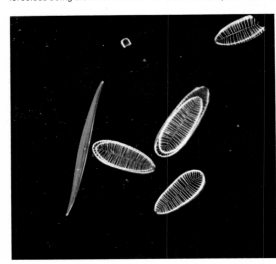

Floating Organisms
The sea offers a great diversity of living conditions. One of these is the possibility of remaining suspended without ever touching the substrate, as is the case with organisms referred to as "plankton". Jellyfish, siphonophores, salps, radiolarians, phytoflagellates, copepods and the larvae of crustaceans and molluscs are just some of the organisms which constitute the plankton. These extremely varied organisms have only one thing in common, the property of floating in the water, either near the surface or in the depths of the sea.

Life in Arctic Waters
During the summer months, the large floating masses of plankton attract a great diversity of animals. The polar waters are actually very rich in animal life and contain a wealth of sponges, bivalve molluscs, crustaceans and echinoderms. The commonest fish species are herrings, sea-bream, cod and salmon, which form dense shoals and provide the standard prey for pinnipeds and cetaceans.

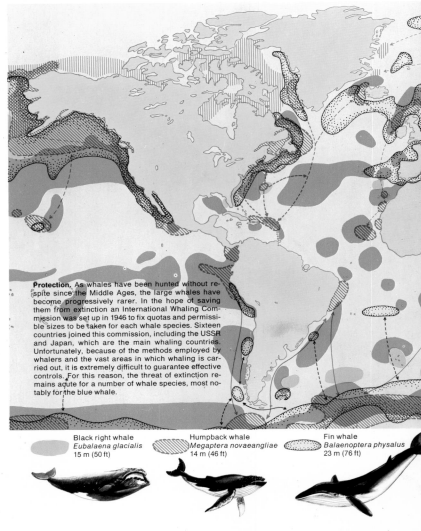

Protection. As whales have been hunted without respite since the Middle Ages, the large whales have become progressively rarer. In the hope of saving them from extinction an International Whaling Commission was set up in 1946 to fix quotas and permissible sizes to be taken for each whale species. Sixteen countries joined this commission, including the USSR and Japan, which are the main whaling countries. Unfortunately, because of the methods employed by whalers and the vast areas in which whaling is carried out, it is extremely difficult to guarantee effective controls. For this reason, the threat of extinction remains acute for a number of whale species, most notably for the blue whale.

Black right whale
Eubalaena glacialis
15 m (50 ft)

Humpback whale
Megaptera novaeangliae
14 m (46 ft)

Fin whale
Balaenoptera physalus
23 m (76 ft)

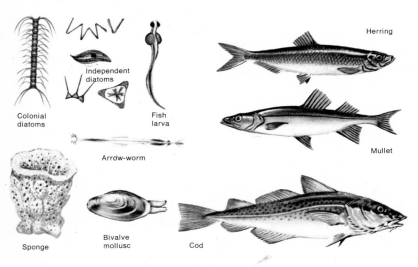

Colonial diatoms

Independent diatoms

Fish larva

Arrow-worm

Sponge

Bivalve mollusc

Cod

Herring

Mullet

Antarctica

Antarctica

Ninety per cent of the ice of the Earth is concentrated in Antarctica, a continent which is almost completely sterile except for occasional tiny patches of land which are free of ice and have a cover of mosses and lichens which provide shelter for herbivorous and carnivorous insects and mites. In contrast to the sterility of the land area, the sea is very rich in living organisms because of its high mineral salt content and the uninterrupted availability of sunlight in the summer, which allows continuous photosynthesis and hence the formation of vast masses of phytoplankton which provide the fundamental food in the Atlantic system. Although the number of species is limited, the number of individuals present actually exceeds that for any other sea in the world. The summit of the food-pyramid is occupied by pinnipeds, cetaceans and a multitude of sea-birds.

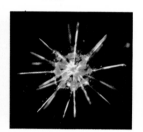

Plankton floats for any of three reasons: 1. The specific gravity is lowered by incorporating bubbles of air or oil, as is the case with the Portuguese man-o-war jellyfish; 2. The body surface is increased to retard sinking, as with the long filamentous appendages of copepods; 3. The performance of swimming movements which maintain the organism at a given position in the water. The illustrations *(right)* show typical members of the plankton: radiolarians, with their characteristic rays of strontium sulphate; copepods; diatoms. The latter, unicellular algae protected by a silicate skeleton, reproduce by fission and they are the most widely distributed plants to be found in the sea.

Antarctic Birds

Antarctica is the main kingdom of sea-birds. Attracted both by the abundance of food and by the isolation of the island coasts, numerous birds move to Antarctica in summer to make their nests: they include rockhoppers *(right)*, sooty albatrosses *(above right)*, grey-headed albatrosses *(above left)*, sea eagles, terns and skuas.

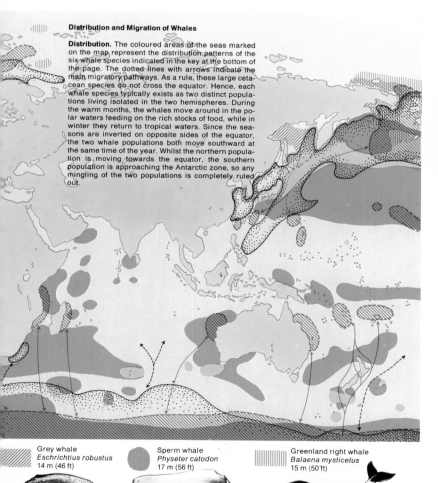

Distribution and Migration of Whales

Distribution. The coloured areas of the seas marked on the map represent the distribution patterns of the six whale species indicated in the key at the bottom of the page. The dotted lines with arrows indicate the main migratory pathways. As a rule, these large cetacean species do not cross the equator. Hence, each whale species typically exists as two distinct populations living isolated in the two hemispheres. During the warm months, the whales move around in the polar waters feeding on the rich stocks of food, while in winter they return to tropical waters. Since the seasons are inverted on opposite sides of the equator, the two whale populations both move southward at the same time of the year. Whilst the northern population is moving towards the equator, the southern population is approaching the Antarctic zone, so any mingling of the two populations is completely ruled out.

Grey whale
Eschrichtius robustus
14 m (46 ft)

Sperm whale
Physeter catodon
17 m (56 ft)

Greenland right whale
Balaena mysticetus
15 m (50 ft)

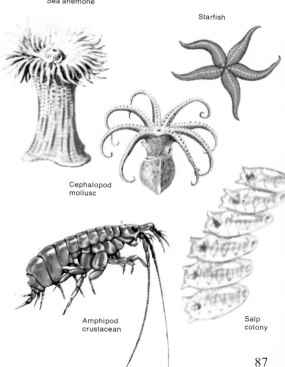

Sea anemone

Starfish

Pinnipeds

Numerous pinniped species breed on the Antarctic pack-ice. Typical visitors are the Weddell seal, the leopard seal and the crabeater seal. The leopard seal is a ferocious predator of penguins. The elephant seal is an endemic species and it is the largest pinniped in the world. It can reach 6.5 metres (21 feet) in length and weigh as much as four tonnes (almost 4 tons). During the breeding season, the males climb onto dry land and engage in fights to stake out territories *(above)*. A month later, they are joined by the females. The illustrations *(right)* show some of the common invertebrates found in Antarctic waters.

Cephalopod mollusc

Amphipod crustacean

Salp colony

Seas and Oceans

The seas of the world constitute a unique system. The sea-water covering the continental shelf regions constitutes what is known as the neritic (or shallow water) zone, whereas the main body of the sea, exceeding 200 metres (660 feet) in depth, constitutes the pelagic (or oceanic) zone. In addition, on the basis of the degree of penetration of sunlight, one can distinguish an upper "euphotic" layer extending down to a depth of 200 metres (660 feet) and beneath that the "aphotic" zone where no light penetrates. Further, it is possible to define two special habitat systems in the transition zone: the intertidal zone and the splash zone.

It is now well established that life began in the sea. More than 3,000 million years ago, the Earth's atmosphere contained nitrogen, ammonia, carbon dioxide, water vapour and a number of other gases, but oxygen was lacking. Ozone, which today forms a protective atmospheric layer filtering out harmful ultraviolet rays, was similarly absent. Hence, until the atmosphere came to contain adequate amounts of oxygen and ozone, life could only develop in the sea, where ultraviolet light could not penetrate. Further, the first living organisms depended upon fermentation rather than on oxidation processes. Gradual increase in the level of oxygen in the atmosphere permitted the emergence of certain forms of aquatic plant-life. The subsequent formation of a layer of ozone in the atmosphere was accompanied by the passage of some organisms from aquatic to terrestrial life. It is now the case that more than eighty per cent of known animal species (approximately one million in all) live on dry land. However, if the insects are excluded (they represent the great majority of land-living animal species) it can be said that sixty-five per cent of the remaining animal species live in the seas.

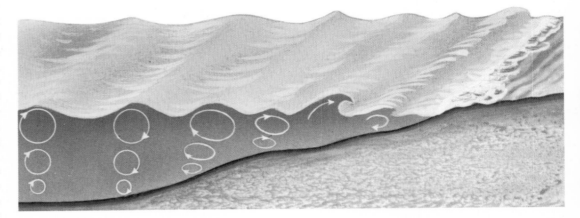

Wave Movements
Tides, currents and waves combine to ensure continuous mixing of sea-water and this in turn leads to the circulation of oxygen, which is indispensable for life under water. Waves, which may reach a height of 10 metres (30 feet) in the open sea, are created by friction between wind and the surface layer of sea water. Water particles are thus forced through a trajectory which becomes elliptical near the coast and produces breakers.

Convection Movements *(diagram of facing page)*
The temperature of sea-water varies both with latitude and with depth. The cold waters of the polar regions tend to move downwards, thus drawing in warm water from more tropical regions, and as a result convection currents between the surface and the depths of the seas are generated.

PENETRATION OF SUNLIGHT

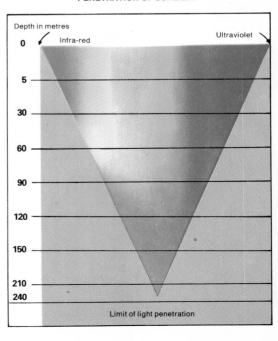

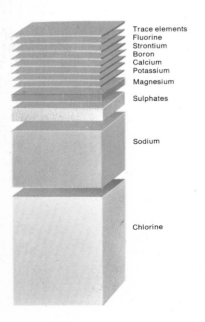

Trace elements
Fluorine
Strontium
Boron
Calcium
Potassium
Magnesium
Sulphates
Sodium
Chlorine

The Salinity of Sea-Water
Seas and oceans together account for more than ninety-seven per cent of the water on the planet. Sea-water contains a number of salts in solution. Although eighty-five per cent of the total salt content is represented by sodium chloride, several other elements are present in significant amounts, in addition to sodium and chlorine. As indicated in the diagram *(left)*, these elements are, in decreasing order of importance: sulphur (in the form of sulphates), magnesium, potassium, calcium, bromine, boron, strontium, fluorine and various trace elements. Gold is actually included among the trace elements, though it occurs only at a very low concentration. Traces of manganese, aluminium and copper are also present. The actual quantity of salt present varies from one place to another, though on average there is 35 gm (1 oz) of salt for every kilogram (2 lb) of water. The salt content obviously declines close to the mouths of large rivers, where there is a considerable influx of fresh water. By contrast, the salinity of sea-water is increased in places where rainfall is limited and there is pronounced evaporation. In the Dead Sea, for instance, 1,000 parts of water contain 200 parts of dissolved salts.

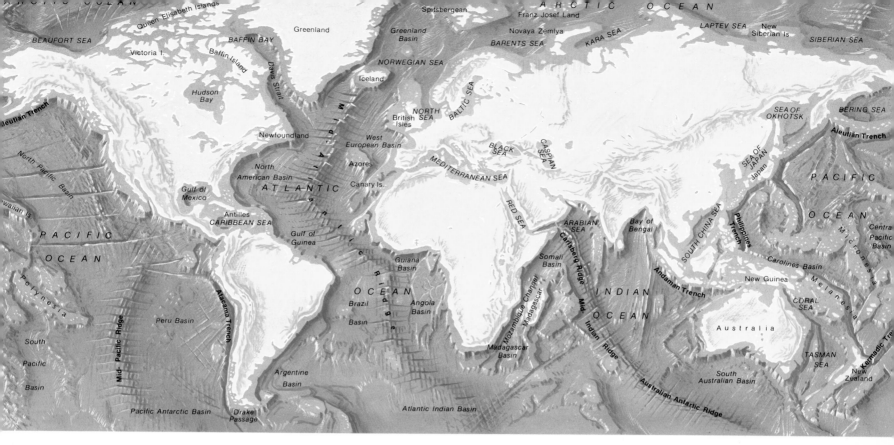

The Oceans

A glance at map of the world is enough to show that "Earth" is not a particularly suitable name for our planet. In fact, 70.8 per cent of the surface of the planet is covered by seas and oceans. Although the sea-water essentially forms a single, continuous body, cartographers have arbitrarily labelled different parts of the seas with special names. In addition to the three main oceans (the Indian Ocean, the Pacific and the Atlantic, which is the largest with a total area of 180 million square kilometres, or 70 million square miles), there are the two polar oceans (Arctic and Antarctic) and a whole variety of seas of different sizes, interconnected to various degrees with the oceans. These seas can be grouped into two categories: 1. enclosed seas, such as the Mediterranean, the Baltic, the Red Sea, the Caribbean Sea and the Sea of Japan; 2. coastal and insular seas, such as the Bering Sea, the China Sea, the English Channel and the North Sea.

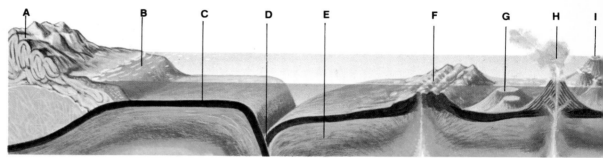

The Ocean Floor

Recent progress in oceanography has revealed the extremely complex conformation of the ocean floor, which was originally believed to be uniformly flat. The diagram *above* provides an idealized cross-section across the ocean floor showing the major features of this still poorly-documented landscape:

A. Continental granitic crust. **B.** Continental shelf. **C.** Oceanic basaltic crust. **D.** Oceanic trench. **E.** Mantle. **F.** Mid-oceanic ridge. **G.** Flat-topped undersea mountain (guyot). **H.** Volcano. **I.** Oceanic island.

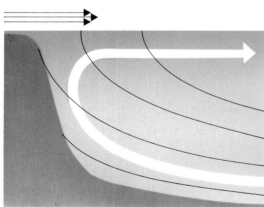

Life in the Seas

Life is present at all levels in the seas, even in the dark abyssal depths, though there is obviously considerable variation in the densities of living organisms. Animals depend upon the presence of plant forms, which represent the initial links in any food-chain. The plants, in turn, depend upon the availability of solar energy for photosynthesis. As a result, plants are concentrated in the surface layer of the sea, particularly where rising currents carry up nutrient salts from the sea-floor where they have been deposited. Plant-eating animals congregate in such zones and they themselves attract numerous predators, including many sea-bird species such as the Arctic tern. (*Right:* A flock of Arctic terns on the wing). The corpses of dead organisms fall to the sea floor and, as they fall through intermediate levels devoid of sunlight and plants, they provide the only source of food for animal species living in the dark. Most benthic species (those living in the depths of the sea on the sea-floor) obtain their food by filtration of the water. Anything which remains is attacked by bacteria and mineralized.

89

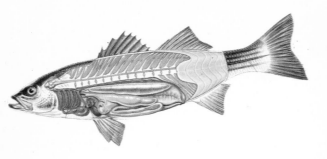

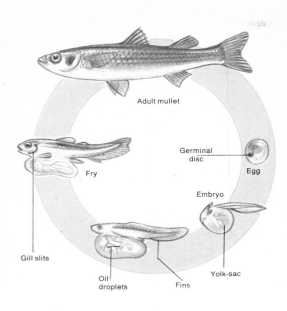

Bony Fish
Fish species such as trout, salmon and sardines all belong to the group known as "teleosts" (bony fish), which are characterized by possession of a lateral line system. *Below:* 1. Sensory cells; 2. Mucus-filled canal; 3. Nerve; 4. Scale. This system has a series of pores leading to nerve-endings which permit the fish to obtain information about the chemical composition and temperature of the surrounding water.

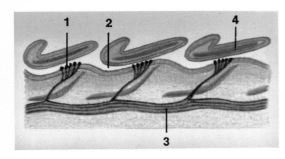

A Zone of Special Adaptations
The seas and oceans are fringed by beaches, lagoons and rocky coasts which provide magnificent scenery. A number of different animal communities can be distinguished in such areas and among them they exhibit some striking adaptations which permit them to live even under the most unusual conditions. Tides, which vary in their extent according to the geographical location, are the major factor influencing life on the coasts. Many animals which inhabit the intertidal zone between the high- and low-tide limits must withstand alternation between submergence and exposure. The plants and animals concerned have developed special adaptations which permit them to survive the adverse phase of the tidal cycle. For example, barnacles (like the one shown *above* attached to the shell of a mussel) withdraw into their shells at low tide. Within their shells, which are solidly anchored to the substrate, they can maintain the moisture that they require. Limpets and periwinkles behave in the same fashion, while other species, such as sea-anemones and tube-worms *(below)*, attach themselves to a hard substrate and remain in the pools which are formed at low tide. *Bottom of the page: A cuttlefish.*

Life in the Water
Birds, mammals, amphibians, reptiles and fish together form the large animal group known as gnathostomes (vertebrates with a proper jaw mechanism). They contrast with the agnathans, which lack a jaw mechanism and which include the cyclostomes (e.g. lamprey), the closest relatives of the gnathostomes. The diagram *above* shows the reproductive cycle of a typical bony fish (the mullet).

The Sea-Horse
The sea-horse is an unusual bony fish with short, wide pectoral fins and a single dorsal fin, but no tail-fin. After fertilization, the female deposits her eggs in the male's ventral sac, where they develop until hatching takes place.

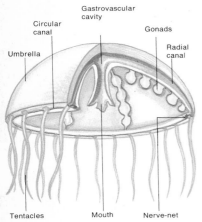

Jellyfish
Jellyfish (or scyphozoans) are coelenterates which maintain themselves upright in the water, with their tentacles hanging below, thanks to special organs (statocysts) which provide the animals with sensory "information" about their positions.

Medusae
The classification of the coelenterates, which includes corals and the jellyfish, is based upon the reproductive cycle. An individual may either be attached to the substrate and reproduce asexually by budding or be a free-swimming "medusa" which can reproduce sexually. In the full cycle, medusae produce either eggs or sperm which fuse in the water and eventually yield larvae (planulae). After spending some time in a planktonic phase, the larva will attach itself to the substrate and produce a polyp. Not all coelenterates pass through the full cycle, however, and three distinct groups can be recognized: 1. Hydrozoans, which live in colonies and exhibit the full reproductive cycle; 2. Scyphozoans (large jelly-fish), which do not exhibit a colonial phase and show budding of polyps in a pattern known as "strobilation"; 3. Anthozoans (most corals and the sea anemones), which do not pass through a medusa phase.

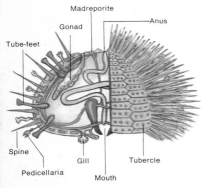

Echinoderms
Echinoderms, such as the sea-urchin, have a complete digestive tube with both mouth and anus. In the sea-urchin, the mouth is associated with a system of five jaws and five teeth known as "Aristotle's lantern", which is embedded in the general body-cavity.

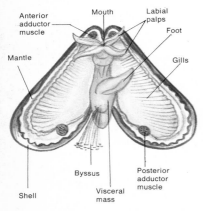

Molluscs
The shell of a mollusc is composed of three layers: 1. An external covering of pigmented material; 2. Calcium carbonate crystallized as prisms (calcite); 3. Calcium carbonate crystallized as plates (aragonite).

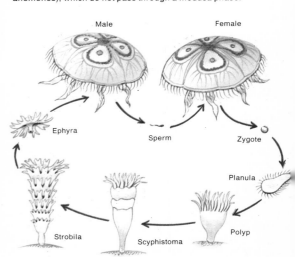

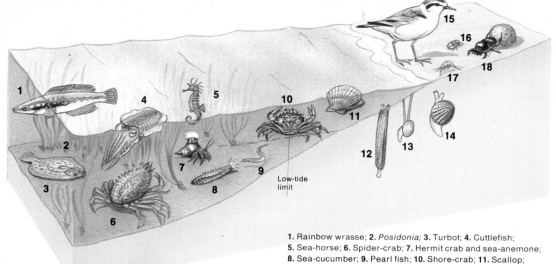

Sandy Shores
Shores often provide a meeting-ground for terrestrial and marine species, since many birds and mammals are attracted by the presence of large numbers of aquatic invertebrates. In tropical regions, monkeys, jackals and wild pigs visit beaches to hunt for shrimps and crabs, which are also sought out by a variety of predatory birds. Along the coasts of Alaska and northern Japan, there are sea-otters to be found which are adapted to aquatic life to such an extent that they will spend the night sleeping on mats of kelp. By day, they feed on limpets, oysters and mussels, which they break open with the aid of a stone "anvil". Two main types of beach are recognized—sandy shores and rocky shores—and they differ considerably in the conditions available for living organisms. On sandy coastlines, the lack of homogeneity in the substrate generally obliges animals to take refuge on the sea-bed, which is the home of crustaceans, worms and bivalve molluscs, all burrowing forms. The diagram (left) shows some examples of animals which occur along sandy shores.

1. Rainbow wrasse; **2.** *Posidonia*; **3.** Turbot; **4.** Cuttlefish;
5. Sea-horse; **6.** Spider-crab; **7.** Hermit crab and sea-anemone;
8. Sea-cucumber; **9.** Pearl fish; **10.** Shore-crab; **11.** Scallop;
12. Razor-shell; **13.** Sunset shell; **14.** Cockle; **15.** Plover;
16. Shore beetle; **17.** Sand-hopper; **18.** Shore dung beetle.

Shore-living Birds
The rich and varied array of bird species which nest on rocky outcrops and small islands represents the peak of the food-pyramid of this habitat. Sea-eagles, gulls (left), cormorants, gannets and guillemots feed on fish and various invertebrates which are caught with widely different techniques. The photograph (right) shows a group of spiny lobsters moving in single file to deeper water, where breeding will take place.

Rocky Shores
On rocky coasts the abundant supply of sunlight and oxygen favours the proliferation of living organisms, although wave movement—which is more pronounced than on sandy shores—creates serious obstacles for the survival of both plants and animals. Three different zones are generally recognized: the splash zone, the intertidal zone and the sublittoral zone. The latter is represented by the relatively shallow, well-lit waters lying above the continental shelf. At the upper limit of the intertidal zone there are blue-green algae and encrusting lichens, among which isopod crustaceans and periwinkles take refuge. Close to the low-tide limit, the algae (sea-weeds) become more abundant, providing food and refuge for a large variety of living organisms such as amphipod crustaceans, sea-squirts, worms and the eggs and larvae of a variety of marine species. The intertidal zone is, above all, the home of barnacles, mussels, sea-urchins, coelenterates (sea-anemones, jelly-fish, etc.), crabs and shrimps. The fish inhabiting the sublittoral zone are generally rather small in size and they are commonly territorial in habit. Gobies lay their eggs in small crevices among the rocks, and the male cleans a circular area around the nest which he defends against intruders. Blennies take refuge in the rock crevices and in empty mollusc shells, which they leave only to forage for food. Some species, such as mullet and sea-bream, have mouths adapted for feeding on molluscs attached to rocks or to sea-weed. Special teeth located in the pharynx are used to break up the shells, which are subsequently fragmented further by a cornified layer on the stomach lining. Pipefish and sea-horses draw in water through their siphon-shaped mouths and thus filter out small marine organisms. Large predators are not residents of this zone, but are occasional visitors attracted by the local abundance of prey. The scheme right illustrates a typical fauna of a rocky coastline.

1. Kittiwake; **2.** Cormorant; **3.** Gannet; **4.** Oystercatcher; **5.** Goose barnacles;
6. Dogfish; **7.** Sea-lily; **8.** Brittle-star; **9.** Lobster; **10.** Corals; **11.** Starfish;
12. Octopus; **13.** Ormer; **14.** Sea-urchin; **15.** Scorpion-fish; **16.** Moray eel;
17. Sea-anemones; **18.** Grouper; **19.** Mussels; **20.** Periwinkles and barnacles;
21. Rock louse.

The Continental Shelf

The waters lying above the continental shelf, which are rich in mineral salts and allow sunlight to penetrate to the sea-floor, can be practically equated with the translucent zone of the sea. In this zone, which has a maximum depth of 200 metres (660 feet), there is no intermediate level and a pelagic layer can be simply distinguished from the benthic (sea-floor) level. Adaptation to life on the sea-floor has required a long evolutionary history, leading to the emergence of a variety of animal species which now occupy the available "ecological niches". A large part of the sea-floor of the continental shelf is covered with vast carpets of sea-weed, providing an ideal habitat for numerous invertebrates. In the deepest areas there are "meadows" of phanerophytes on sandy substrates, and calcareous algae where the sea-floor is rocky. The sea-floor vegetation provides food and refuge for a wide variety of animals: hydrozoans, gastropod molluscs, star-fish, sea-cucumbers, bivalve molluscs and tube-worms. There are a number of hole-boring species—sponges, certain gastropod molluscs and bivalve molluscs (such as ship-worms which severely damage submerged wood)—which are characteristic inhabitants of the continental shelf. Among the fish found in the benthic zone, some—such as cod and whiting—have retained the typical fish body-form and rely upon their agility to escape from predators. Sea-perch, which are more heavily built and slower in the water, take refuge in rock crevices. Other species, however, have undergone marked modification of their body-form: plaice, flounders, common anglerfish and turbots all have flattened bodies and camouflage patterns so that they can avoid being spotted on the sea-floor.

There are also many cartilaginous fish which have become adapted for life on the sea-bed: dogfish, leopard-sharks, several ray species and torpedo rays belong in this category.

In comparison to the number of benthic (bottom-living) species, the number of pelagic fish species is relatively limited. Some species form very large shoals, which apparently provide an effective means of disorienting an attacking predator. However, certain predatory fish, such as barracudas, also live in schools, and some zoologists therefore believe that the advantage of shoal-formation lies in reduction of energy expenditure. Further, such formation of large groups could also have advantages in reproduction.

The Fish
Two major groups of fish are found in the seas: cartilaginous fish (chondrichthyans) and bony fish (teleosts). The first group includes the sharks and the rays, which are characterized by a ventrally-placed mouth and externally apparent gill-slits, by a vertebral column which supports one of the lobes of the tail-fin, by possession of a cartilaginous skeleton and by the presence of placoid, toothed scales. They also exhibit an osmotic balance between their blood and the sea-water which results in retention of part of the urea produced in the circulating blood. In contrast, teleost fish such as salmon and herring are characterized by a forward-pointing mouth, protected gills covered by an operculum (with water entering through the mouth and leaving from behind the operculum), by exclusion of the vertebral column from the tail-fin, by bony scales covered in skin, by possession of a lateral line system *(see page 90)* and by the presence of a bony skeleton.

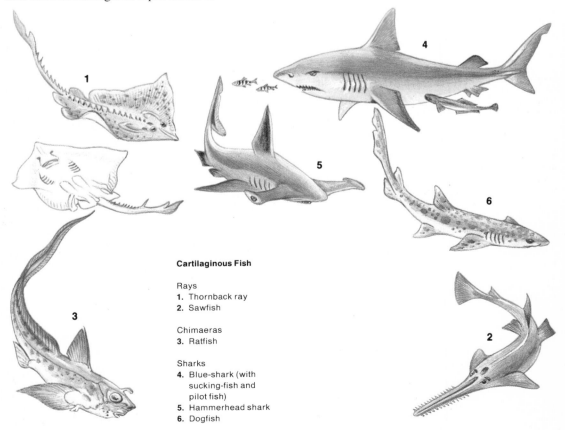

Cartilaginous Fish

Rays
1. Thornback ray
2. Sawfish

Chimaeras
3. Ratfish

Sharks
4. Blue-shark (with sucking-fish and pilot fish)
5. Hammerhead shark
6. Dogfish

A

B

C   D

Life in the Open Sea

The complete absence of natural refuges constitutes the dominant feature of life in the open sea. Animals are nevertheless present everywhere. Away from the continental shelf, sea organisms tend to concentrate in places where rising currents carry nutrient salts from the sea-bed, where they have been deposited, up to the surface. The salts encourage the development of marine plants. With increasing depth in the sea, there is a progressive decrease in penetration by sunlight until complete darkness prevails and plant-life becomes quite impossible. In such abyssal regions, and in the intermediate layer of the sea above them, food is rare and limited to any detritus which falls to the sea-floor. The temperature is also quite low, while pressure is increased. At 1,000 metres (3,300 feet) the pressure is 105 kg./sq./cm. (36 lb./sq./in.), while at 10,000 metres (33,000 feet) it exceeds one tonne/sq.cm. (approaching one ton/sq.in.). These marked changes with depth obviously limit the vertical mobility of marine organisms. As a rule, fish of the abyssal zone are small in size and lacking in bright colours, though they sometimes have small luminous spots constituted by special light-producing cells (photophores). Invertebrates, on the other hand, are often very large in size. For instance, some squids from the abyssal depths which have been washed up on beaches have been found to be up to 17 metres (55 feet) in length. *Above right:* The smooth dogfish, a small shark found in the Red Sea and in the Pacific; *Centre right:* A school of *Seriola dumerii*, migratory fish of subtropical Atlantic waters.

Marine Vegetation

The main plant groups represented in the seas are phytoflagellates and algae (especially seaweeds), ranging from unicellular forms to large colonies. The littoral zone is inhabited by various brown algae and bladder-wracks of the genus *Fucus*, while the sublittoral zone is occupied either by seaweeds of the genus *Laminaria* and by phanerophytes where the substrate is sandy, or by calcareous algae where rocks are present. The schema *(above)* illustrates a variety of marine algae:
A. Planktonic forms; **B.** Green seaweed; **C.** Brown seaweed; **D.** Red seaweed.

Luminous Fish

Luminous fish, which live at great depths in the sea, owe their luminescence to the presence of high local concentrations of light-producing bacteria in special organs known as "photophores". It is possible that the fish are able to exert some control over these luminous organs through the use of particular muscles. Light signals seem to serve a variety of functions in the complete darkness of the abyssal depths, such as: attracting prey; permitting the fish to find its way more easily; and allowing males and females of the same species to recognize one another.

1. Gulper fish; 2–3. Lantern fish; 4. Silver bell; 5. Viperfish; 6. Arborescent deep-sea anglerfish; 7–8. Deep-sea anglerfish with simple lures; 9. Stargazer.

Oceanic Islands

There are three different kinds of oceanic islands: continental islands, volcanic (oceanic) islands and coral reef islands. Most continental islands are linked to the mainland by underwater mountain-chains.

The Pacific Ocean is the site of a great deal of volcanic activity, both along the surrounding coastlines and in the depths of the ocean. The Pacific floor is studded with thousands of volcanic peaks, some of which are thousands of metres in height. In a few cases, these volcanic peaks project above the surface of the sea to form an island. At present, the Pacific contains about two thousand volcanic islands (the Marquesas, the Mariannas islands, Samoa, etc.). The largest of these is Hawaii, dominated by two emergent volcanic peaks exceeding 4,500 metres (14,800 feet) in height, with their bases at a depth of some 5,000 metres (16,500 feet) in the water. Hawaii continues to grow because of the activity of two of its numerous craters, which regularly spew out new material. More ancient islands, by contrast, have more rounded contours because they have been subjected to erosion by atmospheric agents and they have progressively sunk because of the phenomenon of bradyseism (slow up-and-down movement of the Earth's crust). In the course of this process, the coral reef gradually becomes separated from the island itself to form an atoll. Different stages of this process can be observed in the Hawaiian archipelago. The southernmost island, Hawaii itself, has high volcanic peaks with sharp outlines and active volcanoes. Towards the northwest, there are islands with lower, more eroded peaks, and finally there is a large cluster of coral islands, sand-banks and shoals extending for more than 1,600 nautical miles.

Life on the Atolls

According to Darwin's original hypothesis, atolls owe their existence to the building activities of coelenterates (notably the corals) and to the phenomenon of bradyseism. In many cases, they are covered with luxuriant vegetation even though they may be very far away from the nearest mainland coast. Wind deposits dust, organic particles and spores, while birds carry seeds, and marine currents transport algae, animal carcases and fruits (such as the coconut, which can float for a long period of time without losing its capacity to germinate). All of these factors operating together account for the rich, tropical vegetation which is found on most of the coral islands of the Pacific Ocean. Colonization of atolls by animal species is also a chance process. Insects and other tiny animals are carried up to great altitudes by winds and transported vast distances before falling back to the ground. Birds carry among their feathers insects and small molluscs. Reptiles and rodents reach coral atolls on mats of vegetation or on floating tree-trunks. Birds, obviously, are the most successful colonizers of such islands because of their ability to fly. The Hawaiian honeycreepers and the Galapagos finches both represent adaptive radiations from a single ancestral form to produce a variety of differently adapted descendants. *Left, from top to bottom:* Screw-pines; Blackgrass; Red King. *Above:* Fern forest.

The Ocean Floor

The schema *(right)* shows an idealized cross-section of the ocean floor. Major features are the shelves which follow the exposed edges of all the continents, the deep oceanic trench and the mid-oceanic ridge which defines the limit between two adjacent tectonic plates.

Coral Islands

The building activity of the animals known under the general name of 'corals' is particularly prominent in the inter-tropical zone. Mention has already been made of the atolls, coral islands of circular or elliptical form which are particularly concentrated in the Pacific and Indian Oceans, and which together constitute the so-called "Indo-Pacific province". In fact, coral-reefs and islands in the strict sense of the term are limited to this province. Here, the structures concerned are entirely derived from progressive accumulation of the calcareous material secreted by the corals. Coral-reefs of a kind are also found along the coasts of the Atlantic, but in this region the coral material is actually built on a foundation of underlying rock.

Atolls

This term refers to coral islands of circular or elliptical form found in tropical seas. Each atoll surrounds a lagoon with the inner shores sloping down gently from the reef into the lagoon, which is always relatively shallow (100 metres, or about 300 feet, in the largest atolls). On the outside, however, the reef falls away sharply and sometimes even vertically. The animals (corals) which construct these atolls can only live in extremely clean water with a moderate salt content. Further, they require temperatures of 25° to 30°C (77° to 86°F) and above, so they do not occur at depths greater than 40 metres (130 feet). The fact that the bases of present atolls are much deeper than this requires some explanation, and a number of theories have been proposed. The figures *below* illustrate Charles Darwin's classic theory of atoll formation.

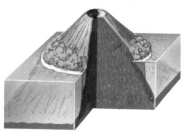

According to this theory, corals initially develop around the base of a volcanic island, which is slowly sinking, to form a coral-reef.

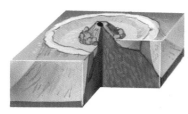

Eventually, the original island has virtually disappeared, while the corals have continued to grow upwards on the foundation already established.

When the island has finally vanished, a circular or elliptical coral atoll is left and vegetation begins to develop on the exposed surface.

Island-Builders

The most important characteristic of coral organisms (anthozoans) is the secretion from their basal disc of skeletal material (coral) containing a mixture of calcium carbonate and iron oxide, which gives the coral its characteristic red colour *(see diagram, left)*. The skeleton has the function of supporting the polyps and the algae which live in a symbiotic relationship with them. Calcium present in sea-water is precipitated at the bases of the polyps thanks to the extraction of carbon dioxide through the photosynthetic processes of the algae, which shift the balance between soluble bicarbonate and insoluble carbonate of calcium.

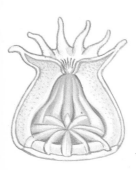

Coral Islands

Atolls, in which sea is combined with dry land to form one of the most enchanting environments to be found anywhere in the world, are confined to the tropical zone between 30° latitude north and 30° latitude south. This is because the calcareous structure of a coral-reef can only be established in sea-water where the temperature remains consistently above 24°C (75°F). Atolls, which are quite common in the Pacific and Indian Oceans, are best developed south of India in the Maldive archipelago. The main trees present are coconut-palms fringing the central lagoon, but here and there are breadfruit trees and various climbing plants. The dense undergrowth is formed by a combination of ferns and a variety of shrubs. A number of different crab species of differing sizes swarm over the beaches, digging tunnels which can be more than 50 centimetres (20 inches) deep. The largest crab species is the robber crab, which climbs up to the top of the coconut-palms and opens the nuts to feed on the pulp. Herons and a number of sea-bird species are natural predators of these crab species. In this unusual habitat there are also a number of rodent species (including the Polynesian rat) and reptiles (including skinks and geckos). In addition, certain marine turtle species visit the shores of atolls to lay their eggs there.

The Colourful World of Coral-Reefs

The largest accumulation of corals in the world, the Great Barrier Reef, stretches over a distance of 1,600 kilometres (1,000 miles) off the north-east coast of Australia. These reefs are formed by vast numbers of coral colonies which provide the ecological foundation for this special habitat. With the passage of time, their calcareous skeletons, which are produced through the coral's special ability to extract calcium carbonate from sea-water, have built up to produce vast mineral accumulations with a variety of form and shape unequalled anywhere else in the world, with the possible exception of primary rainforest. Various species of coral, sea-urchin, starfish, sea cucumbers, sponges, worms, jellyfish, sea-anemones, molluscs and crustaceans have multicoloured patterns and bizarre shapes which are matched by those of a host of fish species. *Centre left:* The rainbow fish *(Anampses meleagrides)*. *Below left:* A butterfly fish *(Jomacanthus octofasciatus)*. The most important predator of this region is the octopus, which is assisted in hunting its prey by its ability to adapt its colour to the background, and by the rapidity with which it can strike. There is another predator which is only concerned with the tiny polyps of the coral-reef, a starfish known as the crown of thorns (genus *Acanthaster*). This predator *(top left)* is very common off the coast of Australia and has caused irreparable damage to the Great Barrier Reef.

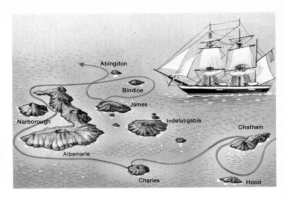

Darwin and the Beagle
The map *(above)* shows the Galapagos Islands and the route followed by the *Beagle*, the vessel that carried Darwin *(left)* on his voyage of exploration. The photograph *(far left)* shows James Bay and the photograph *beneath* shows the general appearance of the vegetation on the shores of these volcanic islands.

The Galapagos

In 1831, Darwin set out on board the *Beagle* with the task of collecting new biological material. In the course of a five-year voyage which took him around the world, he collected a huge mass of information which allowed him to develop the broad basis of his theory of evolution. Study of the large numbers of closely related reptile and bird species in the Galapagos (volcanic islands lying off the coast of Ecuador) led Darwin to conclude that natural selection acted to eliminate individuals which were less able to compete for food, for breeding and so on. Darwin's theory of evolution by natural selection, which is given renewed support whenever the effects of changing environmental conditions are studied, lacked a proper genetic basis and Darwin himself had difficulty in defining the species, which is a fundamental concept for modern biological study.

The Galapagos Animals
During his visit to the Galapagos Islands, Darwin's attention was particularly attracted by the various species of birds and reptiles. The thirteen finch species found on these islands exhibit unique characteristics reflecting the operation of natural selection and they are all descended from a single common ancestor. The main morphological differences between the finch species concern the shape of the beak, which varies according to feeding habits. The most interesting species of all is undoubtedly the woodpecker finch, which removes insects and their larvae from the bark of trees using a cactus spine. A similar pattern of reponse to natural selection is found with the iguana species *(below right)*. Each island has its own characteristic type of iguana with distinctive body characters and behaviour. The pattern of distribution fits well with a hypothesis invoking separate evolution following geographical isolation. The enormous Galapagos tortoises also vary from one island to another, usually showing distinctive features of the shell (carapace). The Galapagos Islands are also visited by a variety of sea-birds which fly there to breed, including the blue-footed booby *(left)* and the frigate bird *(above right)*. The male frigate bird has a large, bright red throat-sac which develops during the breeding season, as a courtship display.

ATLAS
OF PHYSICAL GEOGRAPHY
AND NATIONAL PARKS

EUROPE - PHYSICAL

Scale 1:14 000 000

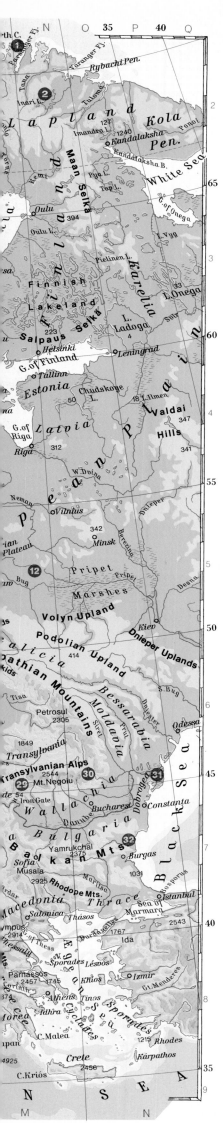

Europe

Europe shows all the characteristic problems of industrialized countries, most notably the consequences of pollution, which pose a serious threat to natural environments. Over the centuries, the original forest covering has been almost entirely destroyed to make way for agricultural development. Irrigation has been introduced on a grand scale, leading to depletion of underground water and hence to a degree of soil dessication. In Southern Europe, exploitation of land over thousands of years has considerably modified the terrain. This is particularly evident in Spain and Portugal, while Italy and Greece have also been severely affected. If the situation is not corrected rapidly, these changes will continue their headlong course and previously fertile regions will become deserts. At the beginning of this century, numerous private organizations were set up in England, Sweden and Germany with the aim of protecting selected sites of particular interest. The first nature reserves of this kind to be established in Europe were located in the Austrian Alps and in the Lüneburg heath region of Germany, a country which now has more than fifty national parks and reserves.

NORWAY

1. Nordkapp (National Park)
The Nordkapp, which is the northernmost projection of the European continent, consists of a sheer granite cliff face which is perfectly smooth and rears 307 m (1007 ft) above the glacial Artic Ocean. The vegetation, which is strictly protected, includes lichens and tiny white flowers. Large colonies of gulls nest on the cliff face and on the large rocky outcrops lying offshore.

FINLAND

2. Lemmenjoki (National Park)
This national park is located in the extreme northern region of Finland, above the Arctic Circle, where the Lemmenjoki River opens into Lake Lokka, which is almost entirely covered with a thick layer of aquatic vegetation and is surrounded by large expanses of quicksand. Vast numbers of insects and birds live in this area.

GREAT BRITAIN

3. Snowdonia (National Park)
The heart of this national park is dominated by Mount Snowdon. The vegetation consists of mixed forest containing both conifers and broad-leaved trees. The animal species that are protected in Snowdonia include wild goats, various mustelids (stoats, weasels etc), voles and a variety of birds such as buzzards, falcons, crows, jays and magpies.

4. Peak District (National Park)
This, the first national park created in England, is characterized by an extremely varied landscape with numerous peaks, valleys and caves.

5. The Brecon Beacons (National Park)
A mountainous area with a particularly rich flora. There is also a great variety of animal species, including many birds.

6. Pembrokeshire Coast (National Park)
This coastland park area is particularly notable for its beautiful landscapes and it serves as an observation station for migratory birds. Thousands of sea birds. notably gannets, cormorants and fulmars, come to nest on the coastal rock outcrops. Grey seals also take refuge along this coast.

BELGIUM

7. Genk (Reserve)
The nature reserve of Bokrijk, which was created for the joint purpose of protecting and studying the plants and animals of this region, lies close to the mining city of Limburg and covers an area of 514 hectares (1337 acres).

THE NETHERLANDS

8. Naardermeer (Reserve)
This is a charming little bird reserve, not far from Amsterdam, which provides shelter for great numbers of waterfowl.

9. Hoge Veluwe (National Park)
Located close to the town of Arnhem, this national park covers a total area of about 6000 hectares (15,000 acres) and contains a mixture of forest and heaths. Deer, moufflon and a variety of other species are able to roam freely.

GERMANY

10. Lüneburger Heide; Lüneburg Heath (National Park)
The extensive heathland of Lüneburg, which lies in the extreme northern region of Germany, is a vast plain covering some 1,135 square kilometres (443 square miles), fringed by marshes and dotted with hillocks which are the remnants of ancient moraines. Only a few shrubs grow on this terrain and it is only on the highest part of the heathland, which constitutes the national park, that a forest has been established after artificial removal of the impermeable surface layer of the soil.

11. Bayerischer Wald; Bavarian Forest (National Park)
This national park was created with the aim of reconstituting the original natural environment composed of a network of marshes, prairies and broad-leaved forests. It is now inhabited by a thriving collection of animal species, many of which were reintroduced: elk, bears, civet cats, storks, lynx, otters and even wolves.

POLAND

12. Bialowieza (National Park)
This region was made into a reserve in 1919 to protect the last remaining European bison and, in 1947, it was made a national park. In addition to bison, it provides refuge for tarpan (small, wild horses), elk, lynx and a large number of bird species.

CZECHOSLOVAKIA

13. Tatra National Park
This is a splendid high montane region where one can see chamois, wolves, brown bears, lynx and marmots.

HUNGARY

14. Valence Lake Reserve
The small lake constituting this reserve is close to the town of Szekesfebervar and closely resembles its large neighbour, Lake Balaton. Half of the surface of the Valence Lake is covered with reeds which provide an ideal natural refuge for a variety of bird species.

AUSTRIA

15. Neusiedlersee and Seewinkel Reserves
These two reserves together encompass a large lake surrounded by huge expanses of reeds which provide a habitat for more than 300 bird species.

ITALY

16. Stelvio (National Park)
This is an alpine site characterized by conifer forests and pastureland. The protected animal species in Stelvio National Park include red deer, roe deer, marmots, eagles and a few brown bears.

17. Abruzzi (National Park)
This very attractive montane park abuts on the snow-capped Montagna della Maiella and Gran Sasso d'Italia. The last surviving chamois of the Abruzzi region are to be found there, along with a small population of brown bears, wolves and eagles.

FRANCE

18. Vanoise (National Park)
This alpine park adjoins the Italian Gran Paradiso National Park which lies on the south-eastern slopes of the mountain. Vanoise National Park contains ibex, chamois, marmots and eagles.

19. Western Pyrenees (National Park)
Created in 1967, this high montane park is adjacent to the Spanish border. It provides refuge for a relict fauna—most notably for brown bears, wild goats and desmans—along with a varied plant community once typical throughout the Pyrenees, including Ramondia, fritillaries, lilies, asters and many other wild flowers.

20. Port Cros (National Park)
This, the only island park including an underwater zone located in the Mediterranean, was established in 1963. Four vegetation zones can be distinguished on the island itself: a coastal zone, a lentisk tree zone, a holm-oak wood zone and a maquis zone. The underwater part of the national park contains Posidonia prairies which are a fundamental part of the local ecological cycle and are severely threatened. Port Cros is also an important staging area for migratory birds.

21. Cévennes (National Park)
This national park, established in 1970, is characterized by a rich and varied flora including the peat-moons and peat-bogs of Mount Lozère, beach groves and natural stands of fir trees. Griffon vultures, black grouse, capercaillie and beavers have been reintroduced to the area. Birds of prey are very common.

22. Ecrins (National Park)
This park was created in 1973 to protect the Haut-Dauphiné massif region. It contains a typical alpine flora, including a number of rare species (e.g. meadowsweet; lady's slipper; alpine columbine). There are also many different insects, 30 mammal species, 90 nesting bird species and 12 reptile species.

23. Mercantour (National Park)
A variety of different sites in Haute-Provence and the Alpes-Maritimes region are included in this park. The vegetation is accordingly quite varied and includes numerous local species such as Saxifraga florulenta. Among the more spectacular animal inhabitants there are chamois, ibex, moufflon and several birds of prey.

24. The Camargue (Regional Park)
Almost the entire area of the Rhône delta is encompassed by this park. There are several hundred bird species there, the most notable being pink flamingoes, herons, kites and eagles. Herds of black cattle and of wild horses add a distinctive character to the region.

SPAIN

25. Valle de Ordesa (National Park)
This montane park lies in the Pyrenees, close to the French border. In addition to numerous bird species, chamois and ibex can be found there.

26. Coto Doñana (Reserve)
Financial assistance from the World Wildlife Fund permitted the establishment of this reserve, which encompasses the Guadalquivir delta and provides the last refuge for the imperial eagle in Spain. The local mammals include fallow deer, wild boar, and a few European lynx and there are a number of reptile species.

YUGOSLAVIA

27. Plitvice Lakes (National Park)
These lakes, lying in the north-western part of Croatia, constitute one of the most beautiful natural landscapes of Yugoslavia. There are 16 lakes, at an altitude of 506 to 603 metres (1660 to 1978 feet), linked to one another by waterfalls and surrounded by superb forests of pine-trees, firs, birches and beech trees. The lake-living animals are particularly abundant, with flourishing populations of trout and crayfish.

28. Mljet (Reserve)
The island of Mljet, one of the many lying off the coast of Dalmatia, extends over 38 kilometres (24 miles) from east to west and has a total area of 98 square kilometres (38 square miles). Its inner coastline marks off several saline lakes which are linked to the sea by narrow channels. The entire island is covered by dense pine forests that sweep up to the summit of Mount Vekj Grad (514 metres; 1686 feet), which is the highest point.

ROMANIA

29. Retezat (National Park)
This national park, which was the first to be created in Romania (1935), lies in the southern Carpathians. The fauna includes colonies of griffon vultures and eagles and a total of 193 species of butterflies and moths (including 4 endemic species), beetles, and other insects. The flora contains 320 plant species, 14 of which are endemic. The pride of the park is the attractive arolla pine.

30. Bucegi (National Park)
Situated in the central Carpathians, this park takes in a limestone massif which has been protected both because of its extremely rare vegetation, including arolla pine, sweet peas and edelweiss, and because of a number of rare animal inhabitants, most notably griffon vultures and eagles. Erosion has carved the limestone into extremely picturesque patterns.

31. Danube Delta
Although this region is neither a reserve nor a national park, it is nevertheless one of the most interesting in Europe because of the waterfowl which live there in their thousands, including a wide variety of species: white pelicans, ibis, egrets, ducks, geese and others.

BULGARIA

32. Kamchiya (Reserve)
South of Varna the Kamchiya River flows into the Black Sea, forming a vast delta. This marshy region covering a total of 5,000 hectares (12,350 acres) is characterized by luxuriant vegetation which is now fully protected: water-lilies and a variety of other aquatic plants, elms, ash-trees, oak-trees and giant lianas.

ASIA - PHYSICAL

The Soviet Union and North Asia

Although human population expansion has not been as spectacular in the Soviet Union as it has been in Europe, the various ecological systems have suffered upheavals which have been, if anything, even more drastic. Artificial diversion of the three great rivers (the Volga, the Dnieper and the Don) which originally discharged their waters into the glacial Arctic Ocean, has given rise to climatic modification, to marsh-formation, and to erosion and deterioration of soils. All of these changes have had serious consequences both for aquatic animals and for land-living mammals.

Hunting, which has been practised intensively for centuries in steppeland and desert areas, has led to the virtual disappearance of certain species. Onagers, gazelles, antelopes and camels are now on the verge of extinction. In an attempt to offset to some extent the continuing destruction of vast regions of natural habitat, more than 7 million hectares (17.5 million acres) of reserves have been created in Siberia. In Japan, by contrast, the problems of human overpopulation have placed great constraints on the dimensions of the 23 national parks which are now present.

1. Astrakhan (Reserve)
The great delta formed by the Volga where it flows into the Caspian Sea has been declared a protected area. It is inhabited by herons, egrets, ibises, wild geese and a large number of migratory bird species.

2. Barsa Kel'mes (Reserve)
This reserve covers a sandy zone extending along the north-west coast of the Aral Sea. It provides a refuge for more than 2,000 bird species when resident and migratory forms are counted together. The reserve also shelters wolves, saigas, goitred gazelles, onagers and jerboas.

South-east Asia and Malaysia

South-east Asia as a whole is affected by the major problem of human overpopulation, which has had the direct effect of forest destruction to make way for agricultural development. Disappearance of forest areas has, in turn, led to a decline in the numbers of many animal species, notably large herbivores which might otherwise have provided a significant food resource. In India, the situation is particularly acute, since there is overpopulation not only among the human inhabitants but also among their domestic animals, with the result that natural resources are disappearing at breakneck speed. Previously fertile regions are turning into deserts, while the rich and varied fauna is disappearing as numerous natural habitats suffer extensive degradation. However, some action has been taken through enlargement of the original natural reserves established during the colonial era and through the creation of new ones.

NEPAL
3. Chitawan (National Park)
Once the private hunting domain of the kings of Nepal, the jungles and grassland areas of Terai are now protected, along with their animal inhabitants: great Indian rhinoceroses, tigers, chitals and other deer, leopards, langurs and gavials. The park can be visited on elephant-back.

INDIA
4. Kaziranga (Reserve)
This zone of forests intermingled with great plains is regularly inundated by the floodwaters of the Brahmaputra. Local animals are Indian rhinoceroses, tigers, water-buffalo, various deer species (including the barasingha), and a wealth of bird species.

5. Keoladeo Ghana (Bird Sanctuary)
This sanctuary, which also goes under the name of Bharatpur, is one of the richest natural reserves for waterfowl, including wild ducks, geese, and cranes which gather on the numerous lagoons in the region.

6. Gir (National Park)
A relatively dry mixed forest and savannah region, this national park provides the last refuge for Indian lions. The park also shelters sambar, nilgai, four-horned antelopes, leopards and a great variety of birds.

7. Kauha (National Park)
This is a vast expanse of forest and grassland which can be seen by visitors riding on elephants. Tigers, swamp deer and a variety of antelope species can all be observed there.

SRI LANKA
8. Wilpattu (National Park)
Located on the western coast of the island of Sri Lanka, this national park contains a mosaic of jungle, grassy plains, swamps and lakes. Elephants, sambar, leopards and a great variety of bird species are to be found in the park.

MALAYSIA
9. Taman Negara (National Park)
This montane park, created in 1938, is characterized by dense rainforest which has remained intact to the present day. Bird species are particularly well represented there, and there are also several notable large mammal species, such as elephants, tigers, gaur, sambar and a few remaining Sumatran rhinoceroses, for which this is one of the last refuges.

BORNEO
10. Kinabalu (National Park)
This park lies in Sabah in the northern part of Borneo, where one can still find extensive areas completely covered with rainforest. Sadly, however, world demand for valuable hardwoods has encouraged systematic tree-felling and many animal species are threatened as a result. The park extends up to an altitude of 4,000 metres (13,000 feet) and includes Mount Kinabalu, from which the park derives its name. Protection is provided for orang-utans, which are now severely menaced with extinction, along with gibbons, various monkey species and probably a small population of the rare Sumatran rhinoceros.

JAVA
11. Udjung Kulon (Reserve)
The island of Java has a large human population and this reserve occupies a relatively small area in the north-west. The reserve, which can be reached only with difficulty, provides a refuge for the extremely rare Javan rhinoceros, for Javan tigers, for leopards and for bantengs.

AFRICA-PHYSICAL

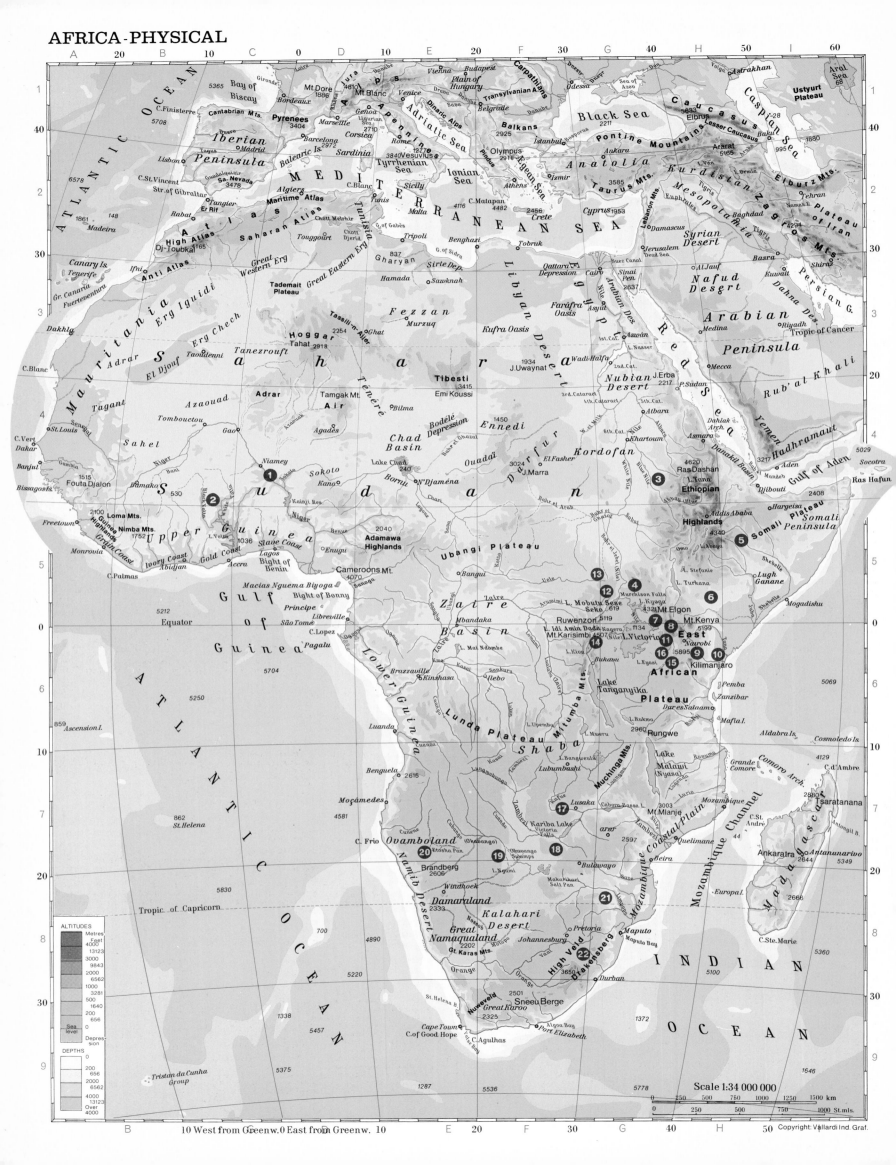

ALTITUDES

Metres	Feet
4000	13123
3000	9843
2000	6562
1000	3281
500	1640
200	656
Sea level	
Depression	

DEPTHS

0	0
200	656
2000	6562
4000	13123
Over 4000	

Scale 1:34 000 000

10 West from Greenw. 0 East from Greenw. 10

Copyright: Vallardi Ind. Graf.

Africa

Up until the end of the last century, Africa was inhabited by enormous numbers of large mammals and several species carried out spectacular migrations at certain times of the year. Explorers who visited the continent during the 19th Century reported herds of zebras and wildebeest moving along in columns stretching over several kilometres. However, the advent of the white man with his firearms marked the beginning of one of the most bloody and wasteful massacres in the history of the animal kingdom. Even North Africa, which was not so richly populated by animals, was subjected to the depredations of white hunters, whose excesses led to the disappearance of Atlas lions and bears along with the Algerian wild ass. Luckily, the colonial authorities realised what was happening before the situation became irreversible, and controls on hunting were introduced. As a result, a number of hunting reserves and natural reserves intended for the protection of animals were created between 1930 and 1950. In fact, South Africa had already developed a policy of protection some decades previously and a number of reserves had been established, with excellent results.

When most of the African states gained their independence in the period 1960–1970, it was widely feared that the accompanying political upheavals might prejudice the protection of animal species by undermining the institutions created for that purpose, which had already begun to yield encouraging results. However, right from the outset, the new African heads of state energetically demonstrated their intention of protecting natural habitats along with their wild animal populations, and they did not shrink from imposing heavy penalties on those who flouted the conservation laws.

In 1968, the heads of several African states were cosignatories to the "African Convention" dealing with the conservation of natural habitat areas. The aim of the agreement was the maintenance of existing protected areas, to be followed eventually by further additions to provide protection for the most characteristic ecosystems. In contrast to the situation in Asia and South Africa, the main problems encountered by African authorities in attempting to carry out conservation projects stem less from overpopulation among the human inhabitants than from traditional customs of land-use. The great herds of wild herbivorous mammals that roam the savannah and Sahel regions and which once provided an inexhaustible reserve of meat and hides, are now regularly decimated by poachers or simply wiped out to make room for domestic animals. The latter, which are consistently maintained in numbers exceeding the natural resources of the soil, rapidly exhaust pastureland and transform it into desert. When this happens, domestic cattle are moved on to recommence the process elsewhere, such that the disastrous consequences for the natural habitat are gradually spread further and further. In addition, certain agricultural practices produce even more serious side-effects and the ever-increasing world demand for wood threatens the existence of forests as well, particularly the montane forests. Since the forests represent a fundamental controlling factor for the accumulation and distribution of rain-water over the entire continent, deforestation leads to erosion and drastic dessication of entire regions. This phenomenon is particularly prevalent in Ethiopia. In other areas, most notably in East Africa, the authorities have understood the importance of protecting the natural environment, even if their standpoint has mainly been determined by economic considerations. In fact, the tourists that come to visit the national parks in large numbers represent a major source of income. Correspondingly, the largest concentration of parks and reserves is to be found in East Africa. In West Africa, there are fewer reserves overall, though this does not mean that they are any the less interesting.

UPPER VOLTA, DAHOMEY AND NIGERIA

1. W-du-Niger (National Park)
This is a vast area of savannah fringed by forests which straddles the borders of these three neighbouring states. The park contains elephants, African buffaloes, roan antelopes, kobs, lions, hippopotamus and crocodiles.

GHANA

2. Mole (Game Reserve)
This large expanse of savannah is inhabited by lions, African buffaloes, roan antelopes, various kob species, oribi, crocodiles and baboons.

SUDAN

3. Dinder (National Park)
This park encloses open savannah grassland of the Upper Nile region and provides protection for cheetahs, leopards, black rhinoceros, elephants, African buffaloes, roan antelopes, giraffes and kudus.

4. Nimule (National Park)
Protection of this region, which has been kept intact, has saved the white rhinoceros from complete extinction. The park also contains elephants, hippopotamus and African buffaloes.

ETHIOPIA

5. Awash (National Park)
The Awash Park is part of a great savannah region. The animals which are protected there include Sommering's gazelle, kobs, greater and lesser kudu, Grevy's zebras and cheetahs.

KENYA

6. Sambura Uaso Nyiro (Game Reserve)
Situated in northern Kenya in a dry area covering a total of 290 square kilometres (98 square miles), this reserve provides protection for a very rich fauna including several monkey species, Grevy's zebras, African buffaloes, black rhinoceros, lions, leopards and crocodiles.

7. Nakuru (National Park)
This zone was declared a national park in order to protect the enormous flocks of flamingos which gather on the shores of Lake Nakuru during the breeding season. Now that the lake waters have been stocked with fish, thousands of pelicans can be found there in addition to a large number of other water-bird species. Waterbuck, baboons and leopards are found around the shores of the lake.

8. Nairobi (National Park)
Of all the national parks in Africa, this one undoubtedly receives the greatest number of visitors. Nairobi National Park has an area of 112 square kilometres (44 square miles) and one can see a great variety of characteristic savannah-living mammals: cheetahs, lions, wildbeest, giraffes, elephants and zebras. The local birds include numerous eagles, notably the crowned hawk eagle and the secretary bird, along with ostriches, pied crows and many other species.

9. Amboseli-Masai (Game Reserve)
The Amboseli Reserve lies at the foot of Kilimanjaro and covers a total area of about 3,200 square kilometres (1,250 square miles). It is relatively easy to visit, since it is only 250 kilometres (150 miles) from Nairobi. The vegetation is largely composed of semi-arid steppe and expanses of savannah with scattered bushes. The reserve contains a large number of animal species, including black rhinoceros, elephants, African buffaloes, lions, leopards, giraffes, zebras and gnus. Birds are also well represented and there are several different weaver-bird species present.

10. Tsavo (National Park)
With its total area of 20,727 square kilometres (8,097 square miles), this is one of the largest national parks in the world. It is a typical savannah zone inhabited by an enormous diversity of animal species, including several antelope species such as the oryx and gerenuk, black rhinoceros, African buffaloes, lions, leopards and—particularly—elephants.

11. Masai Mara (Game Reserve)
This magnificent reserve constitutes a northern extension of the Serengeti National Park of Tanzania, which abuts upon the Kenyan border. Accordingly, one finds most of the species which occur in the Serengeti, such as hunting dogs and cheetahs.

UGANDA

12. Kabalega (National Park)
Just to the north of Lake Albert lies this large national park with an area of about 4,000 square kilometres (1,560 square miles) and at an altitude of 500 to 1,600 metres (1,600 to 5,250 feet). The vegetation in the western part of the park consists of short grass savannah, while in the eastern part there are bushes and small clumps of trees with little foliage. On the banks of the River Nile, whose meandering course is bordered by extensive swamps, papyrus is a dominant feature of the vegetation. The fauna of the park is particularly rich, with elephants, African buffaloes, hippopotamus, giraffes, flamingos, waterbuck, oribi and a considerable variety of other antelope species, lions, leopards, hyaenas, hunting dogs, rhinoceros and crocodiles. The bird life is equally rich, with ibises, geese, fishing hawks, greater crowned cranes, honeyguides and sunbirds.

ZAIRE

13. Garamba (National Park)
This park contains a vast area of dry savannah grassland, once the last refuge of the white rhinoceros in Central Africa. In fact, the rhinoceros has now become so rare in Garamba that it is feared that it might disappear altogether. However, there are thriving populations of chimpanzees, hunting dogs, leopards, lions, hippopotamus and elephants.

14. Virunga Volcanoes (National Park)
A variety of different habitats are present in this park, which covers an extremely large area (8,000 square kilometres, or 3,125 square miles). The grassland areas of the park are inhabited by lions, blesboks, kobs, African buffaloes and hyaenas, while there are large numbers of hippopotamus in the rivers and the dense rainforest provides a refuge for the severely threatened mountain gorilla.

TANZANIA

15. Ngorongoro Crater (Conservation Area)
The protected zone encompassed by this park spans an area of 6,500 square kilometres (2,540 square miles) surrounding the enormous Ngorongoro volcanic crater, which has a diameter of 14 to 19 kilometres (9 to 12 miles) and an internal area of 265 square kilometres (104 square miles). The outer slopes of the crater margin are covered with dense forest, while Lake Magadi is surrounded by large expanses of swampland. The bowl of the crater itself is covered with lush grassland. Of all of the African reserves, this is the one that provides shelter for the greatest number of large mammals: elephants, African buffaloes, hippopotamus, black rhinoceros, leopards, cheetah, various antelope species, giraffes and zebras.

16. Serengeti (National Park)
This extremely extensive park (12,448 square kilometres or 4,863 square miles) provides refuge for a great variety of large mammals, including elephants, lions, black rhinoceros, leopards, African buffaloes, giraffes, immense herds of wildebeest, blesboks, Thomson's and Grant's gazelles, impalas, elands, zebras and kobs.

ZAMBIA

17. Kafue (National Park)
A vast area of 22,400 square kilometres (8,750 square miles) is included in this park. The southern part is largely forested, while grassland predominates in the north. It provides shelter for a great variety of antelopes: roan antelope, oribis, elands, sitatungas, kudus and duikers. There are also large populations of elephants, African buffaloes, hippopotamus, black rhinoceros, lions, leopards, hunting dogs and other mammals.

ZIMBABWE

18. Wankie (National Park)
There is a mosaic of savannah, swampland and forest in this large national park (14,348 square kilometres, or 5,605 square miles). The fauna is correspondingly varied, including elephants, lions, leopards, cheetahs, African buffaloes, black and white rhinoceros, roan antelopes and a variety of other antelope species.

BOTSWANA

19. Moremi (Reserve)
This small reserve is located in the middle of the extensive Okavango swamps, which are inhabited by a considerable number of large mammal species and a large variety of bird species.

SOUTH WEST AFRICA

20. Etosha (National Park)
Lying in the middle of an isolated dry region of South West Africa, this immense national park occupies a total area of 20,700 square kilometres (8,100 square miles). During the rainy season, the entire area is temporarily flooded by the expansion of the great swamp located in the middle of the park. Etosha provides refuge for lions, leopards, cheetahs, rhinoceros, elephants, giraffes, mountain zebras, hyaenas, kudus, impalas and (during the period of flooding) a host of water birds and their natural predators.

SOUTH AFRICA

21. Kruger (National Park)
This is one of the best known and most frequently visited national parks in Africa. It covers a total area of 20,402 square kilometres (7,970 square miles) of savannah which is sparsely wooded only in places. The animals that can be seen there include elephants, hippopotamus, giraffes, lions, leopards and a score of antelope species (notably roan antelopes and kudus). There are also white rhinoceros, which were reintroduced into the park after becoming extinct there at the end of the last century.

22. Umfolozi (Game Reserve)
This game reserve contains both black and white rhinoceros, along with a great variety of antelope species including nyala. The reserve is distinctive in that it is one of the few protected areas of Africa where visitors are allowed to move around on foot.

NORTH AMERICA - PHYSICAL

Scale 1:34000000

North America

The United States of America have the dubious distinction of holding the world record for pollution. Although the human population represents only two per cent of the world total, the U.S.A. produces more than thirty per cent of the polluents. In addition, there is a continuing accumulation of waste products which in some cases pose almost insoluble problems with respect to their disposal. Pollution of the air, the water and the soil has now reached alarming levels. D.D.T. and other insecticides can now be found in the bodies of organisms living on the sea floor, whilst radioactivity is present at all levels in the water through a combination of underground and underwater weapon testing and the radioactive emissions of nuclear reactors. The single example of D.D.T. is enough to show the seriousness of the situation. This substance, introduced into the U.S.A. in 1942, was initially seen as a miracle product, permitting large-scale extermination of mosquitoes bearing malaria and other diseases. However, just a few years later, a striking reduction in the number of certain water-bird species provoked biologists to carry out detailed research into the matter. The final results were conclusive: D.D.T. leaves behind toxic residues after its use. These residues, transported by rivers into the sea or deposited on land in rainwater have been distributed virtually throughout the world. Traces of them have been found in the bodies of seals, penguins and other polar animals. Further, these insecticide residues become increasingly concentrated, and hence increasingly toxic, as they are passed on through the links of a food-chain. Eventually, they reach lethal levels, as was the case with the water-birds, which were the final link in the chain.

Nuclear power-stations, which were designed to replace coal, gas and petrol (all in short supply and all major sources of pollution themselves), have introduced new kinds of ecological disruption which have yet to be fully documented. In addition to the radioactive waste products that they generate, they also give rise to thermal pollution: water from lakes or rivers which is used to cool nuclear reactors is returned to its source at a temperature several degrees higher than normal. This temperature difference is enough to destroy microorganisms which constitute the first link in many food-chains. This, in turn, leads to reduced fish populations and declines in other aquatic organisms and there is an associated increase in the toxicity of certain substances, leading to the spread of diseases.

Nevertheless, the U.S.A. also has the distinction of having created the first national park, that of Yellowstone, in 1872, with the aim of protecting the bison, which was being exterminated on a massive scale to facilitate the construction of the railroad network. Such massacres contributed to the fame of Buffalo Bill, who boasted of having killed more than 5,000 bison. But it was Buffalo Bill himself who realized the gravity of what was happening, and approached the U.S. Government to secure measures to protect the few remaining bison before they became extinct. There are now 280 national parks in the U.S.A., along with 300 sanctuaries and reserves, giving a total protected area of more than eleven million square kilometres (four million square miles). In fact, American public opinion has repeatedly backed compaigns to carry out conservation measures.

Canada, the great neighbour of the U.S.A. has followed this example, and the entire North American continent now has a fairly satisfactory collection of reserves. Only in Mexico are there acute difficulties, largely because of human overpopulation.

CANADA

1. Jasper (National Park)
The fauna of this park includes grizzly bears, mule-deer, bighorn sheep, mountain goats, caribou and elk.

2. Wood Buffalo (National Park)
This is one of the largest national parks in the world. Although it is situated in a region characterized by plains, it is difficult to reach. It provides refuge for herds of bison, elk, wapiti, grizzly bears, black bears, wolves, porcupines, Canadian lynx, wolverines and a variety of other species.

U.S.A.

3. Mount McKinley (National Park), Alaska
This park is located in a region with an extremely rich fauna, including large herds of caribou, Dall sheep, elks, grizzly bears, wolverines, wolves, marmots, pikas, porcupines and (during the breeding season) large flocks of birds which migrate inland from the coast.

4. Olympic (National Park), Washington
The mountains of the Olympic peninsula are exposed to the strong, humid winds from the Pacific Ocean and the resulting heavy rainfall has encouraged the development of dense forest. The entire spectrum of plant forms is represented, ranging from mosses, lichens and fungi right through to trees which can reach heights of 100 metres (300 feet) or more and can live for several hundred years. Between the ice-capped peaks and glaciers of the mountain tops and the forests on the lower slopes lie expanses of pastureland. The fauna is characterized by herds of wapiti, which carry out spectacular migrations during the summer, black-tailed deer, black bears, pumas, Rocky mountain goats and marmots. Five seal species live along the coasts in addition to sea-lions and, further out, whales. Thousands of birds nest on the small islands and rocky outcrops just offshore.

5. Mount Rainier (National Park), Washington
This park is dominated by the volcanic cone of Mount Rainier, whose enormous glaciers feed the surrounding rivers and lakes. In the valleys at the foot of the volcano, the extremely wet climate has favoured the development of dense forests of giant conifers such as the Douglas fir, western red cedar and sitka spruce. The main attraction of the park is the possibility of seeing the glaciers and their effects. In addition, the park provides refuge for 50 mammal species including mountain goats, marmots, wapiti, deer, raccoons and pumas, along with about 130 bird species.

6. Glacier (National Park), Montana
This extremely wild region is characterized by strikingly beautiful landscapes. Animals that can be seen there include bighorn sheep, mountain goats, porcupines, beavers, marmots, elk, wapiti, grizzly bears and black bears.

7. Crater Lake (National Park), Oregon
Crater Lake, the deepest lake of the U.S.A., is surrounded by a patchwork of forests and prairies. This park is now the last refuge for a number of species which have become extremely rare elsewhere: American badger, coyote, bobcat, a few pumas, yellow-bellied marmots and ground squirrels. Numerous bird species are present, including the baldheaded eagle, the golden eagle, the peregrine falcon and a variety of crow and gull species.

8. Grand Teton (National Park), Wyoming
This park is located in the Rockies and the local fauna is typified by grizzly bears, wapiti, mule-deer, marmots and a variety of bird species.

9. Yellowstone (National Park), Wyoming
Created in 1872, this is the oldest national park in the world. It is also the largest national park in the U.S.A., with a total area of more than 2,221,000 acres (898,800 hectares). The high plateau of Yellowstone is located in the heart of the Rockies. Vast conifer forests alternate with deep canyons, lakes and prairies. One of the main attractions of the park is constituted by numerous geysers and hot springs spouting from the earth. The local fauna includes grizzly bears, black bears, elk, bison, wapiti, pronghorn antelopes, bighorn sheep, wildcats, lynx, pumas and coyotes.

10. Wind Cave (National Park), South Dakota
A vast expanse of prairie covering some 28,000 acres (11,300 hectares) is enclosed in this national park. The vegetation is mixed, including North American oaks, conifers, yuccas and cactuses. The most noticeable animals are bison, pronghorn antelopes and prairie-dogs.

11. Isle Royal (National Park), Wisconsin
This national park embraces an archipelago of more than 200 islands in Lake Superior and the natural habitat has been maintained absolutely intact. The rich animal life on the islands includes beavers, muskrats, weasels, varying hares, squirrels, red foxes, bald-headed eagles, fishing hawks, woodpeckers and gulls. There are also some large mammals such as elk which only appeared on the island in about 1900, doubtless during a winter when the great lake was frozen as the islands lie about 86 kilometres (54 miles) from the lake shore. Some twenty years later, wolves also appeared on the islands.

12. Lassen Volcanic (National Park), California
This park, which lies in the north-eastern part of the state of California where the Sierra Nevada meets the Cascade Range, is dominated by a volcanic crater which last showed signs of activity at the beginning of this century. Even today, there are fumaroles (holes emitting gases) and boiling mud springs on its slopes.

13. Yosemite (National Park), California
This is a magnificent canyon with conifer forests cloaking its walls, inhabited by about 80 different mammal species.

14. Sequoia-Kings Canyon (National Park), California
This is a combination of two parks which together cover a distance of 104 kilometres (65 miles) from north to south. The Sequoia Park was established in 1890 to protect the giant sequoia trees which grow on the slopes of the Sierra Nevada at altitudes of 1,000 to 3,000 metres (3,000 to 9,000 feet). The undergrowth is particularly rich in plant species and both willows and hazels are common. Kings Canyon National Park, on the other hand, was created in 1940 to ensure the protection of the canyon site. The canyon is very beautiful indeed and its sheer rocky walls tower up to a height of more than 3,000 metres (9,000 feet). The outer flanks of the canyon are dotted with peaks and channelled by numerous gorges. Glacial lakes and relatively undisturbed forests are to be found there.

15. Grand Canyon (National Park), Arizona
Part of the Colorado Canyon is included in this national park, such that some 100 kilometres (63 miles) of its total length of 350 kilometres (220 miles) are provided with protection. The rocky walls of the canyon, which have been sculpted by water over thousands and thousands of years, provide breathtaking landscapes. Different climatic zones are present in the park. The northern part of the canyon, which has a cool, humid climate, is covered with dense forest, while the southern part is arid. The animals inhabiting these two parts of the canyon are accordingly very different.

16. Bryce Canyon (National Park), Utah
This is a splendid canyon, with numerous rock pillars with the most spectacular shapes. These pillars have been given a variety of picturesque names: Queen's Castle, Gulliver's Hat, Hindu Temple, Wall Street, etc. The park has a total area of more than 36,000 acres (14,600 hectares).

17. Rocky Mountain (National Park), Colorado
This montane park has an area of more than 25,000 acres (10,000 hectares). The undergrowth of the forests covering the slopes of the mountains is characterized by a great diversity of plant species. The emblem of the park is the bighorn sheep, an extremely robust and agile animal which is able to live at considerable altitudes in the mountains. Other notable animal inhabitants are wapiti, beavers, eagles and falcons.

18. Big Bend (National Park), Texas
Situated 482 kilometres (300 miles) south of El Paso, this park covers a total area of 708,221 acres (286,600 hectares) and it is dominated by the Chisos peaks. The region is characterized by desert conditions and it is furrowed by canyons with tall pillars, carved out by the Rio Bravo. Originally, the whole area was covered by the sea, and this explains the presence of sedimentary rocks containing fossils of molluscs and other marine organisms. The flora of this national park includes more than a thousand different plant species. The animals are also numerous, with rabbits, pronghorn antelopes, collared peccaries, mule-deer, the rare kit fox, coyotes, lizards, snakes, tarantulas and more than 300 bird species.

19. Shenandoah (National Park), Virginia
The high plateau on which Shenandoah National Park is situated is part of the Appalachians. The Indian name of the park means "daughter of the stars". It covers an area of 194,000 acres (78,500 hectares) and the landscape is characterized by mixed forests and vast areas of pastureland. This is a favourite spot for bird watchers, since it is possible to spot more than 200 different bird species in a single day.

20. Everglades (National Park), Florida
This park encompasses one of the largest and richest swampland areas of North America, though access to it is somewhat difficult. The Everglades provide a home for countless water bird species, including roseate spoonbills, a variety of birds of prey, pumas, black bears, white-tailed deer and alligators.

21. Mesa Verde (National Park), Colorado
Located in the south-western part of Colorado, this park embraces a plateau 15 square miles (38 square kilometres) in area, lying at an altitude of about 700 metres (2,300 feet). On one side, the plateau is sculpted by deep canyons, while on the top of the plateau itself there are several villages consisting of dwellings carved into the rock which were apparently abandoned about two hundred years before the arrival of Christopher Columbus in America. The principal attraction of this park in fact resides in the atmosphere of mystery surrounding it. Archaeologists have found extremely interesting traces of American prehistory at this site. The Indians that lived on the Mesa Verde for more than eight hundred years actually exhibited many resemblances to the modern Pueblo Indians of the Rio Grande in New Mexico and to the Hopi of North Arizona. However, a great deal remains to be discovered since only a few of the several hundred villages have been fully explored to date.

SOUTH AMERICA - PHYSICAL

Scale 1:34000000

ALTITUDES
Metres Feet
5000 16404
4000 13123
3000 9843
2000 6562
1000 3281
500   1640
200   656
Sea level 0
Depression

DEPTHS
0
200   656
4000 6562
4000 13123
More than

Copyright: Vallardi Ind. Graf.

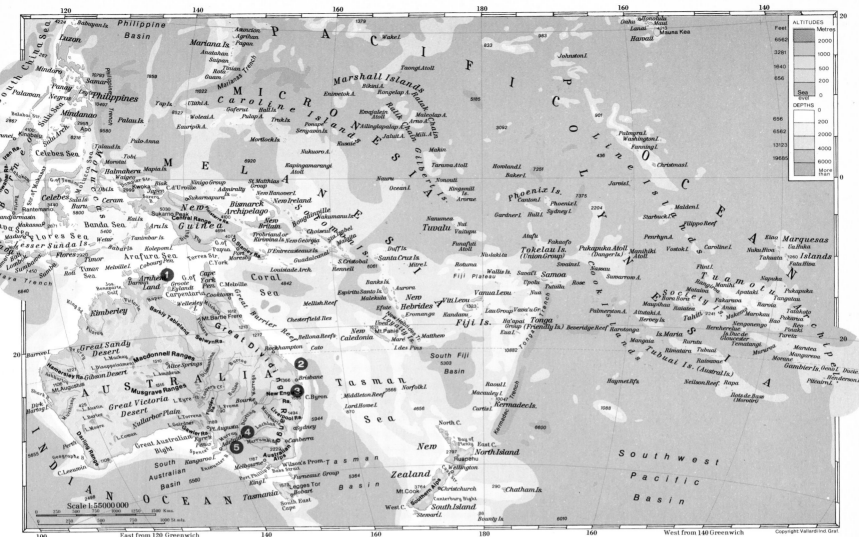

Central and South America

Rapid increase in the human population, particularly in Central America, has been accompanied by an equally rapid decline in the natural environment. Forests are disappearing and erosion is transforming vast areas into sterile wasteland. The fauna, although it is one of the richest and most diverse of the entire planet, is also gradually disappearing while the authorities remain either unable or unwilling to intervene. Luckily, large areas of South America are still uninhabited or difficult to reach, and this, at least for the time being, has to some extent spared them from destruction.

COLOMBIA

1. Tayrona (National Park)
This park stretches along the coasts of the Caribbean Sea and it includes both a marine zone, characterized by coral reefs and numerous tropical fish, and a montane zone, inhabited by a variety of monkey and bird species. Three different marine turtle species come to lay their eggs on the shores.

VENEZUELA

2. Henri Pittier (National Park)
This park, which extends along the Venezuelan coast, has an extensive forest cover. This is one of the most beautiful bird reserves in the world, with more than 500 species recorded. Monkeys and sloths are also common.

ECUADOR

3. Galapagos (National Park)
The introduction of domestic animals such as goats and pigs has unfortunately disrupted the natural ecological balance of these islands, whose flora and fauna inspired Darwin to develop his theory of the origin of species. The main attractions of the islands are still the giant tortoises, the marine and land living iguanas and the Galapagos finches, though other notable inhabitants are albatrosses and a variety of other sea birds.

PERU

4. Pampas Galeras (Reserve)
This reserve was created with the aim of protecting the last remaining vicuñas.

BRAZIL

5. Emas (National Park)
This is a large national park characterized by grassland resembling African savannah, with a few sparsely distributed trees in most areas, except where dense forests line the banks of rivers. The animals found there include maned wolves, various armadillo species (including the giant armadillo), the giant otter (now extremely rare), the giant anteater and the pampa deer. The local birds include rheas and king vultures.

6. Iguaçu (National Park)
This park surrounds the magnificent waterfalls of the Iguaçu River, which carve their way through the subtropical vegetation of the region. Local animals include howler monkeys, capybaras, jaguars and pumas in addition to a great variety of bird species.

Australia and New Zealand

Australia, together with New Guinea and New Zealand, constitutes the only continent in the world which is not plagued by human overpopulation. The first settlers arrived there from Europe in 1788, though the relatively short period which has elapsed since then has been quite long enough for the white immigrants to wipe out 26 mammal species and 13 bird species. Destruction of forests and the extermination of certain animal species has been carried out in a systematic fashion, and this continues to the present day. The kangaroo has been the main victim of such attentions, since it is seen as a rival to domestic animals in the grazing of pastureland.

The Great Barrier Reef, the most impressive structure constructed by living organisms in the whole world, is now threatened by drilling. The reserves and national parks of Australia often provide inadequate protection because sheep and cattle feed right up to their boundaries. Further, because of the relatively flat terrain, which is lacking in natural barriers of any major significance, animals are able to move over long distances, thus escaping proper surveillance and systematic protection. There are now a score of reserves in Australia and about a dozen national parks in New Zealand.

As far as the protection of the oceans is concerned, the most serious problem is that of "neutral" waters which are exploited indiscriminately by a number of different countries. Despite the existence of international regulations, which are often difficult to enforce, several species of fish, whales and seals have already disappeared.

AUSTRALIA

1. Arnhemland (Reserve)
This reserve, situated in the north of Australia, contains a flora and fauna comparable to those of New Guinea. A large number of tropical bird species find protection there, along with freshwater and estuarine crocodiles.

2. Heron Island (National Park)
This national park consists of a tiny fraction of the Great Barrier Reef, providing shelter for a large number of marine animals, some of which have spectacular shapes and colour patterns. There is a variety of different coral formations and the island also provides refuge for sea turtles and various birds.

3. Lamington (National Park)
This is a region of subtropical forest inhabited by a great variety of bird species and some of the most interesting marsupials, including opossums and wallabies.

4. Hattah Lakes (National Park)
In this region, there is a mosaic of forests, lakes and eucalyptus spinneys. Kangaroos, both red and grey, and numerous bird species can be observed there.

5. Wyperfeld (National Park)
This is a zone of brush and open prairies inhabited by kangaroos, mound fowl, parakeets and emus.

Index